Studies in Computational Intelligence

Volume 718

Series editor

Janusz Kacprzyk, Polish Academy of Sciences, Warsaw, Poland
e-mail: kacprzyk@ibspan.waw.pl

About this Series

The series "Studies in Computational Intelligence" (SCI) publishes new developments and advances in the various areas of computational intelligence—quickly and with a high quality. The intent is to cover the theory, applications, and design methods of computational intelligence, as embedded in the fields of engineering, computer science, physics and life sciences, as well as the methodologies behind them. The series contains monographs, lecture notes and edited volumes in computational intelligence spanning the areas of neural networks, connectionist systems, genetic algorithms, evolutionary computation, artificial intelligence, cellular automata, self-organizing systems, soft computing, fuzzy systems, and hybrid intelligent systems. Of particular value to both the contributors and the readership are the short publication timeframe and the worldwide distribution, which enable both wide and rapid dissemination of research output.

More information about this series at http://www.springer.com/series/7092

Álvaro Rocha · Luís Paulo Reis
Editors

Developments and Advances in Intelligent Systems and Applications

 Springer

Editors
Álvaro Rocha
DEI
University of Coimbra
Coimbra
Portugal

Luís Paulo Reis
DSI
University of Minho
Guimarães
Portugal

ISSN 1860-949X   ISSN 1860-9503   (electronic)
Studies in Computational Intelligence
ISBN 978-3-319-86519-5   ISBN 978-3-319-58965-7   (eBook)
DOI 10.1007/978-3-319-58965-7

Printed on acid-free paper

This Springer imprint is published by Springer Nature
The registered company is Springer International Publishing AG
The registered company address is: Gewerbestrasse 11, 6330 Cham, Switzerland

Preface

Editors' Note/Introduction

The development of information systems and technologies and of intelligent systems techniques and methods enabled their application in various industries and businesses processes and services. Assisted by information and communication technologies, automated activities, and decisions methodologies in the organizations, these systems may reduce costs and increase efficiency and reliability because of enhanced control, integration, and optimization of goods production and service delivery. However, the organizational environment as well as the globalized economic environment is equally dynamic and harsh, which poses extra constraints to management and planning and the integration of these systems in organizations focused on the global market. Such constraints demand for intelligent information systems with appropriate design, analysis, and able to satisfy the requirements of various stakeholders with an interdisciplinary approach.

This book is mainly devoted to intelligent information systems (IIS) and to the integration of artificial intelligence, intelligent systems and technologies, database technologies, and information systems methodologies to create the next generation of information systems. These new types of information systems embody knowledge that allows them to exhibit intelligent behavior. This enables them to cooperate with users and other systems in tasks such as problem solving and retrieval and manipulation of a large variety of multimedia information and knowledge. These systems perform tasks such as knowledge-directed inference to discover knowledge from very large data collections and provide cooperative support to users in complex data analysis. IIS are also concerned with searching, accessing, retrieving, storing, and treating large collections of multimedia information and knowledge and also with integrating information and knowledge from multiple heterogeneous sources.

This book includes original and state-of-the-art research concerning theoretical and practical advances on IIS, system architectures, tools and techniques, and successful experiences in intelligent information systems. The book is organized as an interdisciplinary forum in which scientists and professionals could share their research results and report in new developments and advances in intelligent information systems and technologies and related areas.

The book includes 19 chapters, which are extended and improved versions of selected papers from CISTI'2016—11th Iberian Conference on Information Systems and Technologies, held in Gran Canaria Island, Spain, June 15–18, 2016. The chapters describe innovative research work, on several areas of information systems and technologies and applications, with emphasis on intelligent information systems. The chapters are focused on areas such as classification and image analysis, qualitative approaches and qualitative data analysis software, optimization and operational research methods and extensions, Web-based decision systems, verbal decision analysis, agile documentation and ontologies, business process modeling and pervasive business intelligence, intelligent urban mobility, ubiquitous systems, and cognitive learning, among several other topics.

We would like to thank all the contributing authors, as well as the members of the conference coordinating, organizing and scientific committees for their hard and valuable work that assured the high scientific standard of the conference and enabled us to edit this book.

Coimbra, Portugal   Álvaro Rocha
Guimarães, Portugal   Luís Paulo Reis

Contents

Advanced Classification of Remote Sensing High Resolution Imagery. An Application for the Management of Natural Resources

Edurne Ibarrola-Ulzurrun, Javier Marcello and Consuelo Gonzalo-Martin

Abstract In the last decades, there has been a decline in ecosystems natural resources. The objective of the study is to develop advanced image processing techniques applied to high resolution remote sensing imagery for the ecosystem conservation. The study area is focused in three ecosystems from The Canary Islands, Teide National Park, Maspalomas Natural Reserve and Corralejo and Islote de Lobos Natural Park. Different pre-processing steps have been applied in order to acquire high quality imagery. After an extensive analysis and evaluation of pan-sharpening techniques, Weighted Wavelet '*à trous*' through Fractal Dimension Maps, in Teide and Maspalomas scenes, and Fast Intensity Hue Saturation, in Corralejo scene, are used, then, a RPC (Rational Polymodal Coefficients) model performs the orthorectification and finally, the atmospheric correction is carried out by the 6S algorithm. The final step is to generate marine and terrestrial thematic products using advanced classification techniques for the management of natural resources. Accurate thematic maps have already been obtained in Teide National Park. A comparative study of both pixel-based and object-based (OBIA) approaches was carried out, obtaining the most accurate thematic maps in both of them using Support Vector Machine classifier.

Keywords Remote sensing · Natural resources · Image pre-processing · High resolution image · Pansharpening · Orthorectification · Atmospheric correction · Classification · OBIA

E. Ibarrola-Ulzurrun (✉) · J. Marcello
Instituto de Oceanografía y Cambio Global, IOCAG, Universidad
de Las Palmas de Gran Canaria, ULPGC, Las Palmas de Gran Canaria, Spain
e-mail: edurne.ibarrola101@alu.ulpgc.es

E. Ibarrola-Ulzurrun · C. Gonzalo-Martin
Facultad de Informática, Departamento de Arquitectura y Tecnología de Sistemas
Informáticos, Universidad Politécnica de Madrid, Madrid, Spain

© Springer International Publishing AG 2018
Á. Rocha and L.P. Reis (eds.), *Developments and Advances in Intelligent Systems and Applications*, Studies in Computational Intelligence 718,
DOI 10.1007/978-3-319-58965-7_1

1   Introduction

Ecosystems provide a wide variety of useful resources or services that enhance the human welfare, however, in last decades there has been a decline of them [1]. Development of image processing techniques, applied to new remote sensing imagery to obtain accurate and systematic information, is associated with the monitoring of these marine and terrestrial ecosystem services. Thus, it is important to develop reliable software methodologies for the analysis, conservation and management of these resources.

Regarding coastal ecosystems, they are the most complex, dynamic and productive systems in the world [2] and it exists a difficulty in the remote monitoring of sea parameters because the low reflectivity of these covers. In this context, the monitoring and evaluation of the quality of water surface is fundamental to the management of their resources.

Concerning terrestrial ecosystems, most forest and vegetation maps are produced under supervision and mostly from the fieldwork and the use of aerial photographs or orthophotos. Thus, the challenge is the generation of a robust, automatic, and cost-effective methodology for the systematic mapping of endemic plant species and colonizing plants using remote sensing techniques. However, there exists some weaknesses in the classification and analysis of plant and tree species by remote sensing such as the lack of reliability of the maps, the high cost of revamps, the lack of automation and computerization, etc. Moreover, the images of remote sensing satellites need to be processed to generate useful products for the management of natural resources.

Remote sensors provide raw data images, hence, it is necessary to apply correction techniques and perform image pre-processing in order to obtain high quality spectral and spatial imagery. Most optical sensors provide a multispectral and a panchromatic image of the same scene, simultaneously recorded. By fusion techniques at pixel level (pansharpening), the spatial resolution of the multispectral image can be improved by incorporating information from the panchromatic image.

In the last decades, various algorithms have been developed [3–8], mainly based on the Discrete Wavelet Transform [9, 10] and on the Modulation Transfer Function [11], as well as improvements in traditional algorithms [12], however specifically algorithms for new multispectral sensor must be created. In addition, acquired images from high resolution sensors require the correction of the topographic relief effects as it decreases or increases the radiance illuminated depending on the slope of the areas. The main algorithms are not fully validated [13–15]. In [16] is presented a more comprehensive analysis, but only for medium-resolution imagery. Furthermore, radiometric corrections include the elimination of certain errors and the conversion of digital values to radiance levels. Then, atmospheric correction algorithms transform the TOA (Top of Atmosphere) radiance into ground reflectance removing the atmosphere absorption and scattering effects. Different methods have been implemented such as DOS, 6S, ATCOR, FLAASH and QUAC. Recently, there have been several comparative studies [17, 18].

Once applied these techniques, image processing is needed in order to generate useful marine and land high level products. The main algorithms are directed to the generation of multispectral indices or classification and segmentation techniques for obtaining a thematic map which shows clearly the parameter under the study. The state-of-art in classification techniques is oriented towards the implementation of more efficient algorithms such as Support Vector Machines and fuzzy classification and other novel techniques such as those based on spatial context, texture, multi-temporal information, etc. Due to the high spatial resolution imagery, it is necessary to highlight new analysis technologies as OBIA (Object Based Image Analysis) [19–21].

In this context, the general objective of this work is the development of advanced processing techniques which serve to obtain accurate information for the conservation of natural resources when they are applied to very high resolution remote sensing imagery. Thus, novel techniques are implemented in each of the processing stages. Techniques for the image fusion, segmentation and classification of multispectral bands will be developed as well as algorithms for atmospheric and topographic correction.

Objectives at more specific level are:

• Determine the necessary data for the characterization of ecosystems to be analyzed.
• Apply novel pre-processing algorithms adapted to the features of images selected to provide spatial and spectral information of high quality.
• Develop specific products for the management of terrestrial and plant resources.
• Develop specific products for the management of coastal and marine resources.
• Conduct studies of the state of conservation of natural resources.

2 Methodology

The first stage of the work includes three steps: (1) gather all the information (images and data) required in each area of study, (2) characterize of each ecosystem and (3) validate the products generated by remote sensing satellite imagery.

The study is focused in vulnerable ecosystems found in different islands of the Macaronesia region since it is considered both geologic and biodiversity hotspot due to their volcanic origin. Thus, three areas in Canary Islands, Spain (Fig. 1) were chosen as a representative sample of these heterogenic ecosystems having the availability of obtaining field data. Such areas are: Teide National Park located in Tenerife Island, Maspalomas Natural Reserve in south Gran Canaria Island and two areas of north Fuerteventura, Corralejo Natural Park and Islote de Lobos.

WorldView-2 (WV-2) orthoready imagery of the three ecosystems have been employed in the study (Fig. 2). It provides very high spatial (Panchromatic: 0.46 m, Multispectral: 1.84 m) and spectral (8 Multispectral bands) resolution data.

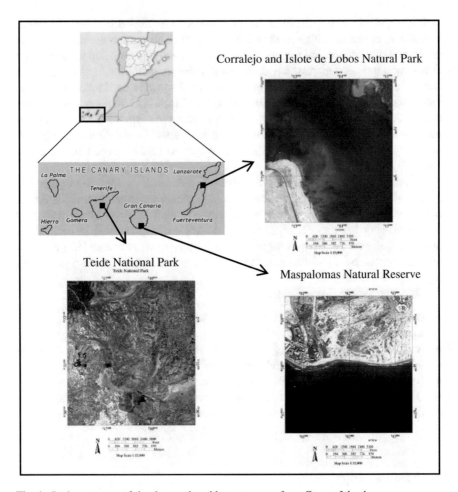

Fig. 1 Study area map of the three vulnerable ecosystems from Canary Islands

In situ available measurements to study the natural resources such as water quality measurements, bathymetry, sea bottom identification, sea and land surface reflectivity and ancillary data (terrain elevation maps, orthophotos, etc.) were obtained and collected the same days the WV-2 satellite captured and image of these areas.

Figure 3 shows a diagram with every step required in order to obtain marine and terrestrial products [22]. After a detailed review of the state-of-art in pansharpening techniques, pansharpening algorithms that could achieve good performance with WV-2 imagery were selected and evaluated in order to assess which technique gave a better fused image for a posteriori classification (Table 1). The algorithms achieving the best performance have been identified by visual and quantitative assessment. By visual evaluation, the main errors at global scale were observed and

(a)

(b)

(c)

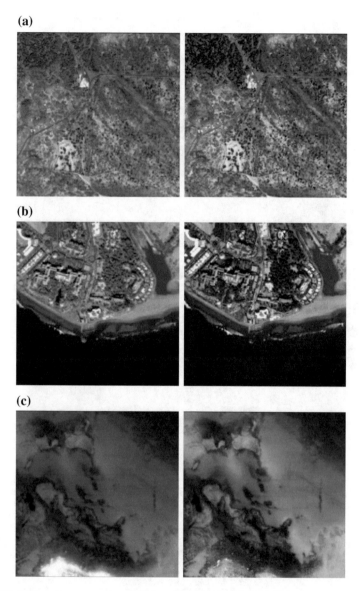

Fig. 2 WorldView-2 scenes analysed (512 × 512 pixels): **a** Teide National Park; **b** Maspalomas Natural Reserve; **c** Corralejo Natural Park and Islote de Lobos

then, local artefacts were precisely analyzed. Besides, a number of statistical evaluation indices were used to measure the fused image quality. All this process is presented in [23].

In addition, orthorectification is carried out using a RPC model which replaces the rigorous sensor model with an approximation of the ground-to-image

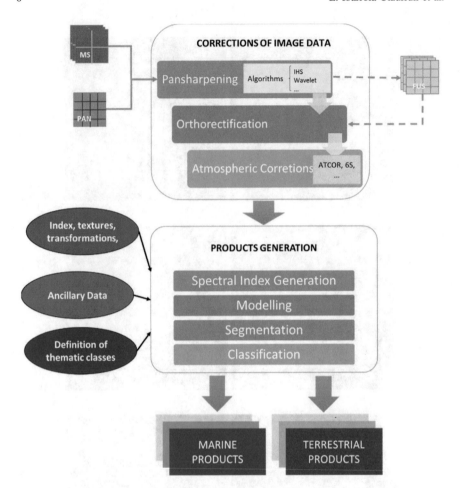

Fig. 3 Diagram of processing methodology. MS: Multispectral image; PAN: Panchromatic image and FUS: Fused image

Table 1 Pansharpening algorithms evaluated	**Pansharpening algorithms**
	Gram—Schmidt
	Fast intensity hue saturation
	Hyperspherical color sharpening
	Modulation transfer function—Generalized Laplacian pyramid
	Modulation transfer function—Generalized Laplacian pyramid —High pass modulation non-matching within PAN and MS bands
	Wavelet 'à trous'
	Weighted wavelet 'à trous' through fractal dimension maps with—window size of 7, 15 and 27

relationship [24]. Finally, a comparative study of 5 atmospheric correction algorithms (FLAASH, ATCOR, 6S, DOS and QUAC) was carried out.

Once images have been corrected and enhanced, the generation of products for the management of coastal and land resources can be carried out.

Regarding land ecosystems, in this study the challenge of detection, classification and analysis of endemic plant species by remote sensing is addressed. Pixel-based and object-based classification techniques are implanted in order to obtain accurate information of the vulnerable ecosystems. The whole study is presented in [25].

Up to this point of the study, accurate thematic maps of Teide National Park have been obtained.

The vegetation classes selected for the Teide National Park ecosystem were set by the National Park conservation managers: *Spartocytisus supranubius* (Teide broom), *Pterophalus lasiospermus* (Rosalillo de cumbre), *Descurania bourgaeana* (Hierba pajonera) and *Pinus canariensis* (Canarian pine). Moreover, urban, road and bare soil classes were included.

The two classifications approaches: pixel-based and object-based classification are carried out using ENVI 5.0 image processing software (Exelis Visual Information Solutions, Inc., a subsidiary of Harris Corporation) [24] and eCognition Developer (Trimble Geospatial) [26] which is based on an object oriented approach to image analysis. eCognition is optimized for cost-effective OBIA classification of very high resolution imagery.

The classification algorithms analyzed were, at pixel-based approach, *Maximum Likelihood Classification, Mahalanobis Distance Classification, Support Vector Machine* and *Example Based Classification* using the *K-Nearest Neighborhood Classification*; and at object-based approach, the *Bayes Classification, Nearest Neighborhood Classification, K-Nearest Neighborhood Classification* and *Support Vector Machine*. The statistical accuracy assessment technique used is the Error Matrix which reports two accuracy measures, the Overall Accuracy and the Overall Kappa accuracy.

In case of marine ecosystems, high resolution satellite sensors increase the use of remote sensing techniques to assess water quality. However, estimating the parameters of water quality is a complex process. Thus, we plan to develop validated and operational algorithms to study such parameters. On the other hand, generating maps of bathymetry using high resolution images is an innovate field. In the study new multiband algorithms based on a physical model of radiative transfer for the marine bathymetry to about 25 m depth with high resolution sensors will be generated. Finally, advanced classification techniques are applied to generate coastal seafloor maps using OBIA techniques combined with robust pixel-based classification (Support Vector Machine) and using all available spatial and spectral information.

To finish, once validated the availably of the products associated with the different marine and terrestrial resources, the researches will perform a detailed report on the conservation status of each natural resource analysis.

3 Results

As regards the pansharpening results, it is important to highlight the study of the pansharpening techniques not only in land and urban areas but a novel analysis in areas of shallow water ecosystems i.e. Maspalomas Natural Reserve and Corralejo and Islote de Lobos Natural Park. The best fused technique for each study area is shown in Table 2 and Fig. 4. Although the quality of the fusion algorithms does not show a high difference in land and coastal image, coastal ecosystems, as Corralejo, require simpler algorithms, such as Fast Intensity Hue Saturation, whereas more heterogeneous ecosystems, i.e. Teide and Maspalomas ecosystems, need advanced algorithms, as Weighted Wavelet '*à trous*' through Fractal Dimension Maps.

Once, the best fused image was obtained, orthorectification was carried out. The orthorectification errors in each scene were compared visually with images obtained from GRAFCAN and their geodesic points (http://visor.grafcan.es/visorweb/) (Fig. 5 and Table 3).

Afterwards, the atmospheric correction is carried out and the assessment has revealed that 6S is the most suitable for the scenes (Fig. 6) regarding different settings.

Finally, classification techniques were carried out at pixel based and at object based approach in Teide National Park image. Figure 7 shows the most accurate thematic maps obtained with both approach, which it is achieve in both cases with Support Vector Machine classifier (Radial Basis Function; Gamma in Kernel Function: 0.5; Penalty Parameter: 200). The Accuracy Assessment Support Vector Machine classifier at pixel based approach is 92.75% of Overall Accuracy and a Kappa coefficient of 0.89, while at object based approach is 89.39% at Overall Accuracy and a Kappa coefficient of 0.85.

As final point, comparing the accuracy results for the pixel-based and object-based approach, it is important to note the good results obtained by OBIA approach. Although the best Overall Accuracy (92.75%) and Kappa coefficient (0.89) is obtained by Support Vector Machine in the pixel-based approach, the computation time is about 12 h while the second best result, Support Vector

Table 2 Best pansharpening algorithms for each scene

Scenes	Fused algorithms
Teide	Weighted wavelet '*à trous*' through fractal dimension maps (window size 7)
Maspalomas	Weighted wavelet '*à trous*' through fractal dimension maps (window size 15)
Corralejo	Fast intensity hue saturation

(a)

(b)

(c)

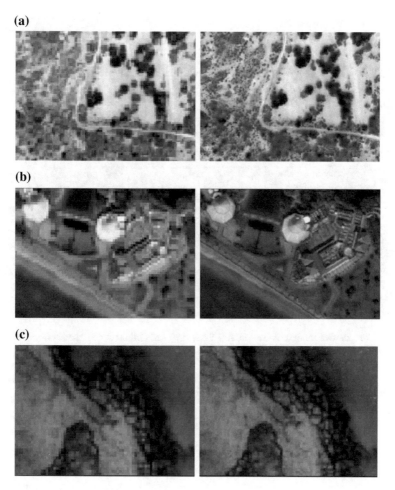

Fig. 4 Original MS and final fused images of each scene zoomed. **a** Teide National Park; **b** Maspalomas Natural Reserve; **c** Corralejo Natural Park

Machine at OBIA approach (Overall Accuracy: 89.39% and Kappa coefficient: 0.85), has a computation time is 39.33 s. This leads to Support Vector Machine at OBIA approach classification applied to the fused image, to be more suitable classifier in order to obtain an accurate thematic map this difficult heterogenic shrubland ecosystems.

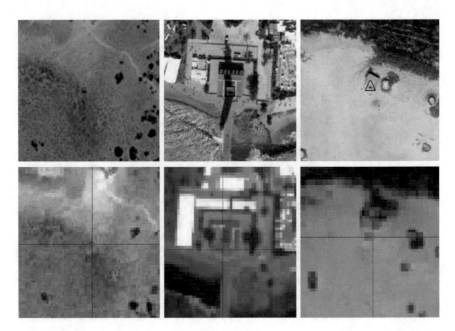

Fig. 5 Comparative images of GRAFCAN Visor (*up*) and Orthorectified images from Teide (*left column*), Maspalomas (*middle*) and Corralejo scenes (*right*)

Table 3 Orthorectification error in the different scenes

Position in orthorectified images	Average error (m)
Teide scene	3.11
Maspalomas scene	8.00
Corralejo scene	9.17

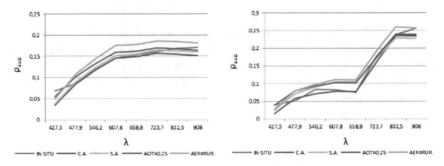

Fig. 6 Reflectivity in the different bands and for the different settings using 6S algorithm in bare soil (*left*) and vegetation (*right*)

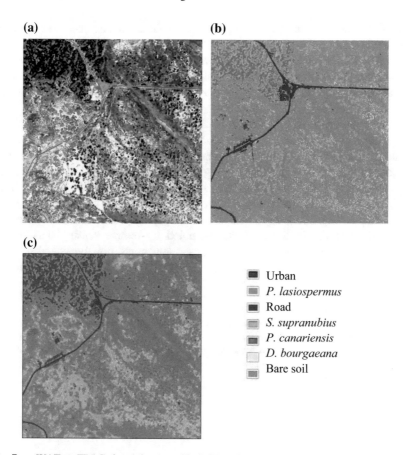

Fig. 7 a WAT ⊗ FRAC fused image with RGB color composition and thematic maps of WAT ⊗ FRAC fused image using Support Vector Machine classifiers **b** at pixel based approach and **c** at object based approach

4 Conclusions

The main goal of the project is to develop advance processing techniques to apply in very high resolution remote sensing imagery, in order to obtain accurate information for the conservation of marine and land natural resources. The final aim leads to the conservation and management of vulnerable ecosystems.

It is observed the necessity to perform accurate pre-processing steps in order to improve the spectral and spatial quality information in the imagery such as image fusion or pansharpening, orthorectification and atmospheric correction. Related to the image fusion results, there are no remarkable differences in the way pansharpening algorithms perform in land (Teide National Park) and coastal (Maspalomas Natural Reserve) areas. Moreover, images with low variability, such as Corralejo Natural Park, requires simpler algorithms i.e. Fast Intensity Hue

Saturation, rather than more complex and heterogeneous ecosystems (Teide National Park).

On the other hand, it is observed as well the importance of the processing steps for achieving an accurate thematic map of ecosystems, for instance in the selection of classes of interest, algorithms parameters setting and the training and testing samples collection.

Thus, future steps will be a more detailed evaluation of both at pixel based and object based classification approaches for land and marine ecosystems. In the case of land ecosystems, we will generate robust, automatic and cost-effective methodology for the systematic mapping of endemic plant species and colonizing plant. Furthermore, it appears a great difficulty when classifying species from coastal ecosystems due to the low reflectivity of water. Thus, we will generate both bathymetry and seafloor maps. Besides, a tool implementation will be carried out which allow access to the products generated by remote sensing data for the management of natural resources in coastal and land areas. Finally, a detailed report on the conservation status of each natural resource analysis will be carried out.

Acknowledgements This research has been supported by the ARTEMISAT (CGL2013-46674-R) project, funded by the Spanish Ministerio de Economía y Competitividad). Moreover, the authors want to acknowledge to the Teide National Park conservation managers (Jose Luis Martín Esquivel y Manuel Marrero Gómez) for defining the classes of interest and examining the results obtained. This work was completed while E. I-U was a Ph.D. student in the IOCAG Doctoral Program in Oceanography and Global Change and funded by the Spanish Ministerio de Economía y Competitividad with a FPI grant (BES-2014-069426).

References

1. Pagiola, S., Von Ritter, K., Bishop, J.: Assessing the economic value of ecosystem conservation. World Bank, Washington, DC. © World Bank. https://www.openknowledge. worldbank.org/handle/10986/18391 License: CC BY 3.0 IGO. (2004)
2. Barange, M., Harris, R.P.: Marine ecosystems and global change. Oxford University Press, Oxford (2010)
3. Amro, I., Mateos, J., Vega, M., Molina, R., Katsaggelos, A.K.: A survey of classical methods and new trends in pansharpening of multispectral images. EURASIP J. Adv. Signal Process. 79 (2011)
4. Fonseca, L., Namikawa, L., Castejon, E., Carvalho, L., Pinho, C., Pagamisse, A.: Image fusion for remote sensing applications. In: Image Fusion and Its Applications. www. intechopen.com (2011)
5. Laben, C.A., Brower, B.V.: Process for enhancing the spatial resolution of multispectral imagery using pan-sharpening. In: Google Patents (2000)
6. Li, X., Li, L., He, M.: A novel pansharpening algorithm for WorldView-2 satellite images. In: International Conference on Industrial and Intelligent Information (ICIII 2012) (2012)
7. Li, X., Qi, W.: An effective pansharpening method for WorldView-2 satellite images. In: IEEE (ed.) International Conference on Estimation, Detection and Information Fusion (ICEDIF 2015) (2015)
8. Padwick, C., Deskevich, M., Pacifici, F., Smallwood, S.: WorldView-2 pan-sharpening. In: American Society for Photogrammetry and Remote Sensing (2010)

9. Amolins, K., Zhang, Y., Dare, P.: Wavelet based image fusion techniques—An introduction, review and comparison. ISPRS J. Photogramm. Remote Sens. **62**(4), 249–263 (2007)
10. Lillo-Saavedra, M., Gonzalo, C.: Spectral or spatial quality for fused satellite imagery? A trade-off solution using the wavelet à trous algorithm. Int. J. Remote Sens. **27**(7), 1453–1464 (2006)
11. Vivone, G., Alparone, L., Chanussot, J., Dalla Mura, M., Garzelli, A., Licciardi, G., Restaino, R., Wald, L.: A critical comparison among pansharpening algorithms. IEEE Trans. Geosci. Remote Sens. **53**(5), 2565–2586 (2015)
12. Tu, T.M., Su, S.C., Shyu, H.C., Huang, P.S.: A new look at IHS-like image fusion methods. Inf. Fusion **2**(3), 177–186 (2001)
13. Richter, R., Kellenberger, T., Kaufmann, H.: Comparison of topographic correction methods. Remote Sens. **1**(3), 184–196 (2009)
14. Gao, B.C., Montes, M.J., Davis, C.O., Goetz, A.F.: Atmospheric correction algorithms for hyperspectral remote sensing data of land and ocean. Remote Sens. Environ. **113**, S17–S24 (2009)
15. Riaño, D., Chuvieco, E., Salas, J., Aguado, I.: Assessment of different topographic corrections in Landsat-TM data for mapping vegetation types. IEEE Trans. Geosci. Remote Sens. **41**(5), 1056–1061 (2003)
16. Hantson, S., Chuvieco, E.: Evaluation of different topographic correction methods for Landsat imagery. Int. J. Appl. Earth Obs. Geoinf. **13**(5), 691–700 (2011)
17. Agrawal, G., Sarup, J., Bhopal, M.: Comparision of QUAC and FLAASH atmospheric correction modules on EO-1 Hyperion data of Sanchi. Int. J. Adv. Eng. Sci. Technol. **4**, 178–186 (2011)
18. Samadzadegan, F., Hossein, S., Pourazar, S., Hasanlou, M.: Comparative study of different atmospheric correction models on WorldView-2 imagery. In: XXIII Congress of the International Society for Photogrammetry and Remote Sensing. Melbourne (2012)
19. Blaschke, T.: Object based image analysis for remote sensing. ISPRS J. Photogramm. Remote Sens. **65**(1), 2–16 (2010)
20. Peña-Barragán, J.M., Ngugi, M.K., Plant, R.E., Six, J.: Object-based crop identification using multiple vegetation indices, textural features and crop phenology. Remote Sens. Environ. **115**(6), 1301–1316 (2011)
21. Garcia-Pedrero, A., Gonzalo-Martin, C., Fonseca-Luengo, D., Lillo-Saavedra, M.: A GEOBIA methodology for fragmented agricultural landscapes. Remote Sens. **7**(1), 767–787 (2015)
22. Ibarrola-Ulzurrun, E., Gonzalo-Martín, C., Marcello-Ruiz, J.: Analysis of land and marine resources by processing high resolution satellite images. In: IEEE (ed.) 2016 11th Iberian Conference on Information Systems and Technologies (CISTI), pp. 1–4 (2016)
23. Ibarrola-Ulzurrun, E., Gonzalo-Martin, C., Marcello-Ruiz, J., Garcia-Pedrero, A., Rodriguez-Esparragon, D.: Fusion of high resolution multispectral imagery in vulnerable coastal and land ecosystems. Sensors, **17**(2), 228 (2017)
24. ENVI: ENVI user's guide. Research System Inc. (2004)
25. Ibarrola-Ulzurrun, E., Gonzalo-Martin, C., Marcello-Ruiz, J.: Influence of pansharpening techniques in obtaining accurate vegetation thematic maps. In SPIE Remote Sensing. International Society for Optics and Photonics, pp. 1000515–1000515 (2016)
26. Baatz, M., Benz, U., Dehghani, S., Heynen, M., Höltje, A., Hofmann, P., Lingenfelder, I., Mimler, M., Sohlbach, M., Weber, M.: eCognition user guide. Definiens Imaging GmbH (2001)

Analyzing the Impact of Strategic Performance Management Systems and Role Ambiguity on Performance: A Qualitative Approach

Enoch Asare and Sue Conger

Abstract This research questions the assumption that strategic performance measurement systems (SPMS) that define strategic goals at the individual job level reduce role ambiguity and ensure desired employee outcomes. Through qualitative research of both white-collar and blue-collar jobs, we seek to determine the types of jobs most amenable to SPMS guidance.

Keywords Strategic performance measurement systems · Role ambiguity · Job related information · Strategic outcomes

1 Introduction

Strategic performance measurement systems are software programs used by businesses to align strategy with individual performance [10, 11, 17]. They have been indirectly related with individual performance through mediators [4–6]. Role ambiguity is an example of these mediators as hypothesized by Asare and Conger [1]. Role ambiguity is the perception that one lacks information required to be proficient on the job, therefore, leading to feelings of helplessness and confusion [1, 8, 9].

SPMS enhance employees' performance through low role ambiguity by providing them with job-relevant information [10]. Job-relevant information is information needed by employees to perform their job duties the extent to which it should be clearly communicated to them [8, 10]. The more job-relevant information available to employees the higher their expected performance and productivity [1]. This is because, job-relevant information improves jobs understanding [8, 10].

E. Asare · S. Conger (✉)
University of Dallas, Irving, TX, USA
e-mail: sconger@udallas.edu

E. Asare
e-mail: easare@udallas.edu

© Springer International Publishing AG 2018
Á. Rocha and L.P. Reis (eds.), *Developments and Advances in Intelligent Systems and Applications*, Studies in Computational Intelligence 718,
DOI 10.1007/978-3-319-58965-7_2

Fig. 1 Proposed research model

While prior research illustrates how SPMS relates to performance, it is largely supported by quantitative, positivist analysis. Also, the phenomena by which role ambiguity affects individual performance is not well understood as there are conflicting findings [11], creating a gap in the literature. With a qualitative study from a constructionist worldview, we seek to bridge this literature gap by exploring the phenomena by which SPMS affects individual performance with the research questions below (Fig. 1):

RQ1: How does an SPMS affect role ambiguity?
RQ2: How does role ambiguity affect individual performance?

This research is important because SPMS are represented as applying to all work; yet, no evaluation of different types of jobs has been conducted. This research will seek to determine if there is a range of jobs to which SPMS apply because it might alter company behavior in SPMS use. In addition, it may show whether further research is needed to clarify all of the contingencies relating to SPMS and its applicability.

This research contributes to the literature by exploring the SPMS—performance phenomenon to provide a deeper understanding of how SPMS and performance relate to each other. Further, we bring a subjective constructionist perspective to the conversation as current research is more tilted towards positivist perspectives where there is one reality. With this goal, we will conduct multiple qualitative case studies to obtain the perspectives of managers and non-managers as well as the perspectives of white and blue collar workers on the SPMS—performance phenomenon.

In the next sections, we detail the relevant literature and research conduct. The next section describes relevant research on SPMS and role ambiguity. Next, SPMS research is linked with role ambiguity and job-related information. Then, role ambiguity links to individual performance are explored. Lastly, the research methodology and expectations are developed.

2 Background

This section provides a summary of research on SPMS, role ambiguity, job performance, and their linkages to identify the gaps that this research seeks to analyze.

2.1 SPMS

An SPMS is "an information system containing financial and non-financial measures that are derived from strategy and designed to align individual actions with
organizational strategy" [3]. Thus, SPMS serve four main purposes—convert
strategy into measurable actions, clarify job duties, monitor individual and
department-level performance, and provide feedback on the status of performance
to the individual and to management [1].

Strategy is a course of action that enables an entity to compete in the market
place [12]. This course of action is translated into levels of detail, such as initiative,
project, and, eventually into individual job targets and performance measures that
can be clarified, monitored, and measured by SPMS [6, 10]. For this purpose,
SPMS are equipped with financial and non-financial performance measures and
feedback tools that communicate and clarify employee job duties [4, 11, 16]. To be
effective in clarifying tasks and monitoring performance SPMS should properly
convert strategy into measurable actions [13].

SPMS clarify tasks by communicating job requirements that have been converted
from the strategy from which it is cascaded down to business unit goals and then to
individual-level goals [10, 11]. Thus, through the decomposition process, managers
define strategic work components and assign them to departments and individuals
across the organization [11]. Moreover, individuals can always go back to the SPMS
to further clarifty their goals. For example, an employee who does not remember
how his/her tasks relate to the organization's strategy can obtain that information
from the SPMS. Those receiving conflicting task directions from different mangers
can also obtain clarity from the SPMS as it documents employe task requirements.

In addition to clarifying and communicating job requirements to departments
and individuals, SPMS also aid managers in monitoring individual-level activities
and achievements [5, 6] and departments [2, 3, 14]. SPMS can monitor financial
and non-financial goals that have been converted from strategy [10]. In particular,
SPMS helps managers to monitor performance by comparing actual individual and
department-level financial and non-financial goals with targets prescribed by the
SPMS [5]. Thus, SPMS enables managers to monitor both individual and
department-level performance.

To summarize, SPMS convert strategy into measurable actions, clarify tasks, and
help managers, and monitor performance at the individual and department levels by
comparing actual performance to expected performance. This research focuses on
the relationship between SPMS and individual-level performance.

2.2 Sources of Role Ambiguity

Role ambiguity is a lack of clarity about job requirements that results in an
employee becoming confused about his/her job [8, 9, 11]. Role ambiguity may be

viewed as the difference between the information needed to perform a job and the information that is actually available [8]. Role ambiguity may come from a variety of sources in an organization [6, 8]. During the strategy setting process top management teams develop strategy and assign them to department heads who then deconstruct the strategy into individual-level goals [6, 10, 11]. However, if the goals are deconstructed incorrectly, assigned to the wrong departments or individuals, or otherwise misconstrued in some way, the errors can result in role ambiguity for the individual assigned to execute the tasks.

An example illustrates the problems that might occur. Assume a hypothetical organization in which the management team designs strategy and asks the planning department to cascade it down to individual-level goals. The cascading strategy used may result in a couple of ambiguities down the organization structure [13]. First, the planning department, responsible for receiving strategy from the management team and cascading it down to various departments may not clearly cascade down the strategy to the departments. This is a source of role ambiguity as the incoherent cascading of strategy to various departments translates to lack of understanding of what is required of individual employees [8, 13]. Second, role ambiguity may come about when department heads receive the correct information from the planning department but do not understand the roles to be played by their departments in implementing the strategy [6, 8]. Third, managers may receive the right strategy information from department heads but may not be able to assign corresponding job duties to employees in clear and coherent manner, leading to employee level role ambiguity [8]. Lastly, role ambiguity may come about when the management team is unable to convey the message behind strategy to managers and department heads in a clear manner [13]. While all these sources of role ambiguity are important, we focus on manager-level and employee-level role ambiguity and refer to them as role ambiguity in this study.

2.3   SPMS and Role Ambiguity

The extant research indirectly associates SPMS with individual performance [4–6]. For example, Burney and Widener [8] assert SPMS that are closely tied to strategy minimize role ambiguity by providing individuals with job-related information and that resulting low role ambiguity enhances individual performance. Despite Burney and Widener's [8] assertions, the phenomenon by which SPMS affect performance through reduced role ambiguity is not clear [11].

SPMS minimize role ambiguity by translating strategy into individual-level goals [8]. To be effective in minimizing role ambiguity, SPMS should be able to convert strategy into measurable actions and communicate those actions to employees in a coherent manner [11]. Also, SPMS should provide employees with feedback as to how they are performing in relation to strategy [11]. SPMS that are able to perform these functions are deemed to be closely tied to strategy and clearly specify employee goals [4, 6, 9, 11]. These SPMS minimize role ambiguity as employees become aware of what is expected of them [8]. Further, SPMS formalize goals and job duties by putting them in writing, which minimizes employee uncertainties about their job duties [8]. Thus, an SPMS that is closely tied to strategy minimizes role ambiguity by converting strategy into actionable tasks, communicating tasks to employees and monitoring tasks to ensure that they are performed according to strategy [9].

Some authors (e.g. [10]) argue that SPMS provide individuals with job-related information as a means of minimizing role ambiguity. Burney and Widener [8] also argue that job-related information serves as a source of reference that help individuals to clarify task ambiguities with managers and guide them in performing their day-to-day duties. Further, individuals who are provided with job-related information are more likely to return to the job-related information source for more information [8]. Thus, while role ambiguity and job-related information seem similar, role ambiguity reflects the extent to which individuals understand their duties while job-related information reflects the extent to which individuals have information to do the understanding [8]. In summary, SPMS increases job-related information, which in turn should decrease role ambiguity.

Despite the arguments in the above paragraphs, three things are still not clear in the literature regarding how an SPMS relates to role ambiguity. First, it is not clear whether job-related information actually has to increase before role ambiguity could be minimized or both happen at the same time. For instance, from the earlier example, delivery drivers know from the SPMS that for every hour they spent delivering products last year, they have 42 min for the current year (30% less). However it is not clear whether the drivers' role ambiguities reduce just because they know their current year goals from the SPMS. This is because there is little to no role ambiguity in truck driving so simply altering drivers' goals does not necessarily alter their role ambiguities. Rather, altering goals by simply reducing prior year delivery times could annoy drivers which can lead to low morale and high turnover among drivers. Thus, research that explores how SPMS affects both white and blue collar role ambiguity is necessary. This study serves this purpose by studying managers, non-managers, white and blue collar workers in a single study. Second, detailed perspectives of individuals (employees and managers) on how they perceive SPMS to minimize role ambiguity with or without job-related information is not clear. Lastly, as far as we are aware only one research [8], has studied how SPMS affects role ambiguity and was conducted from positivist perspectives on management accountants (white collar workers), tilted more towards the researchers' worldviews. Moreover, because the study surveyed only white-collar workers it is not clear as to how SPMS affect role ambiguity of blue-collar workers. Hence, there is the need to explore how SPMS affects both white and blue-collar workers in a single study. With research question 1: How does an SPMS affect role ambiguity? This literature gap will further be bridged by exploring how SPMS minimize role ambiguity from a constructionist perspective where the truth is subjective. In doing this, we will rely on individual (employees and managers) accounts of how in their practical experiences SPMS affect role ambiguity with or without job-related information.

2.4   Role Ambiguity and Individual Performance

When role ambiguity is minimized, it is argued to enhance individual performance [8, 11]. However, this phenomenon is not well understood, as there are conflicting findings in the literature [11]. For example, some authors argue there is a negative relationship between role ambiguity and individual performance, where individual performance increases as role ambiguity decreases [8, 11]. Others assert there is a U-shape relation between individual performance and role ambiguity, where individual performance decreases with increases in role ambiguity to a point beyond which individual performance increases even if role ambiguity increases [11]. Still, others argue there is an ∩-shape distribution between individual performance and role ambiguity, where individual performance increases with increases in role ambiguity to a point beyond which individual performance decreases as role ambiguity increases [11]. Alternatively, we may find that more than one of the distributions hold for different types of jobs.

In sum, while the current streams of research provide insights into how role ambiguity affects performance, they studied limited contexts. Also, these studies mostly are positivist where there is one reality (c.f. [6, 8, 11]). Moreover, few studies provide detailed accounts of employees' views on how role ambiguity affects their performance. Thus, a unifying account of how role ambiguity affects performance that uses a subjective constructivist perspective may provide new insights into the phenomena. Consequently, with research question 2: How does role ambiguity affects individual performance? We employ a case study methodology from a subjective constructivist perspective to explore the phenomenon by which role ambiguity affects individual performance.

3   Methodology

Given the lack of clarity on the SPMS—performance phenomenon and the inductive nature of our research questions [16], this research adopts a multiple qualitative case study methodology. This methodology will enable us to explore SPMS in different contexts [16] with multiple data sources [7]. We will explore the research questions from a subjective constructivist perspective, where the truth is in the eyes of the beholder, to provide research participants the opportunity to share their stories and experiences through semi-structured interviews [7].

Interview data will be analyzed by identifying interpretable units and organizing them into meaningful categories [16, 18]. With this approach, we expect the outcome of this study to be a deeper understanding of the SPMS—Performance phenomenon. We will follow the steps recommended for conducting qualitative case study research and analyzing qualitative data (by Creswell [7]; Saldaña [18]). These are appropriateness of case study for the research under consideration, case identification and selection, data collection and analysis, and case interpretation.

3.1 The Cases

Qualitative case study is suitable for 'how' research questions with a goal of providing an understanding of a phenomenon [7, 15]. Qualitative cases explore real-life, bounded systems through contextual detail and in-depth analyses [7], which is the objective of this research. Further, given conflicting explanations from prior research on the SPMS—performance phenomenon and, especially, how role ambiguity affects performance, multiple qualitative case studies should improve our understanding of the phenomenon. With these objectives, we will conduct separate case studies in five business segments of a utility company located in the Southern United States, including wholesale, retail, mining, generation, and transmission units. We will conduct semi-structured interviews to obtain the perspectives of managers, non-managers, white and blue-collar workers on how SPMS affect their performances. The managers will represent both white-collar and blue-collar units. Hopefully, the distinct characteristics of participants and their experiences with SPMS in each context (case) will not only bring diverse perspectives to the study but also make the study more robust by explaining some of the contradictions of past research [16].

Case studies are appropriate for qualitative research when there are well defined contexts within which the researcher seeks an understanding of a phenomenon [7, 20]. This study meets these criteria as exploring how SPMS affect individual performance within the contexts of the five business segments, each with its own unique management and corporate culture. To gain access to the organization, formal letters of intent explaining the research and the benefits to the organization will be sent to the directors of accounting and human resources, both of whom have agreed to the study in informal conversations.

Cases should bring different perspectives to the phenomenon being studied [7]. To meet this criterion, the cases we have identified for this study have distinct SPMS and job requirements that come with their own role ambiguities, making them appropriate for the study. For each business segment, we will select a total of 18 individuals as recommended by Saldaňa, [18] for initial study. At least three managers and three non-managers as well as six white-collar and six blue-collar workers will be selected. This will enable us to represent all blue and white-collar perspectives of the SPMS—performance phenomenon. We will select participants by drawing alternate random and purposeful samples of individuals such that purposeful samples follow random samples for each organization studied. We will end with three purposeful and two random groupings. Our goal is to select demographics that will be underrepresented after drawing random samples, using purposive sampling. Thus, samples such will mix white-collar and blue-collar workers and managers in each of five business segments.

Qualitative case studies are more credible when cases' characteristics reflect the phenomenon being studied [7, 20]. Given this facts and that this study is intended to provide a more complete understanding of the SPMS—performance phenomenon, the five cases were chosen as having experienced the phenomenon in very different contexts. Thus, the cases provide credibility because of their differences.

Components of the SPMS—Performance phenomenon as illustrated by current literature are the focus of this research. These are:

1. How SPMS relate to role ambiguity and
2. How role ambiguity relates to individual performance.

The first component addresses research question 1 and the second addresses research question 2. In summary, the cases that we have identified and will be selecting for this study are more likely to have characteristics that reflect the SPMS —performance phenomenon and as a result make our study more robust [16].

3.2   Data Collection

Before conducting interviews, we will first collect secondary data. This will be comprised of employee performance evaluation scores for the past 5 years, firm mission and vision statements (as documented on company website), and strategy. In addition, we will document business unit goals and objectives as prescribed by the company's SPMS and how those goals relate to firm strategy. The objective of collecting these secondary data is to enable us document how the company's strategy relates to its vision and mission. It will also inform us on how the strategy has been translated into business segments' and employees' goals and objectives. We will also interview the planning group (the team that translated the organization strategy into individual job duties). This, along with the secondary data will provide us with a better understanding of how, on paper, the SPMS is intended to function. Understanding the data will make us better informed when designing interview protocols to be used in primary data collection [18] and enable us to properly interpret them [16].

In conducting interviews, we will rely on semi-structured techniques as they blend structured and unstructured formats. Semi-structured interviews will enable us to keep participants focused on interview questions and at the same time allow them to elaborate on their feelings, attitudes and perceptions about the SPMS— performance phenomenon [7, 15]. In addition, since we will be collecting similar data across the business segments, we will start interviews as and when business segments are ready to be interviewed. However, we will finish conducting all interviews in one business segment before moving to another. The objective here is to rely on our experience with one segment to anticipate and address potential challenges in subsequent segments. In addition, since most employees schedule their vacations after July, we will collect data from June 1st to July 31st in 2017.

These decisions on data collection meet the criteria that data should be extensive and drawn from multiple sources [7] as multiple sources bring different perspectives (subjective realities) into the study, a basic premise on which constructionist qualitative studies are based [20]. We discuss our interview protocols in the next paragraphs.

3.3 Interview Protocols

Interview protocols will cover topics such as SPMS, role ambiguity, job-related information, and individual performance as highlighted in research questions 1 and 2. We will also leave room for participants to extend the discussions to cover issues that may arise [7, 15]. Moreover, we will encourage respondents to interpret and answer questions from their own perspectives [7, 20]. We intend to allow participants to answer questions from their experiences with SPMS and expect that the order of questioning may deviate from the protocol as we encourage participants to express themselves freely [16].

Since we will be interviewing 15 individuals for each business unit, we plan to conduct five sets of face-to-face interviews in each business unit where three different people will be interviewed in each set. To encourage participants to give honest answers, we will interview managers and non-managers separately. We expect each interview section to last between 1 and 2 h. All interviews will be recorded and transcribed afterwards for analysis to prevent data loss [16]. To ensure that we are consistent with the themes that we will gather from interviews, we will compare and validate the interview data with the secondary data [7, 15, 18].

As interviews are completed, we will conduct reflection sessions to discuss what we have observed in the session. We will also discuss our own reactions to the interviewees to ensure that personal perspectives do not bias research results. Questions of interpretation will be revisited with an individual respondents.

Once all interview data have been collected, we will discuss transcribed data between ourselves on how SPMS' relate to role ambiguity (research question 1) and how role ambiguity in turn relates to individual performance (research question 2) [7, 18]. These discussions will be based on how each of us perceive and interpret interview results [16, 18]. Findings will be validated with key informants from the interviewees.

3.4 Data Analysis

Within-case and between-case analyses will be conducted and will include case descriptions, histories, and chronologies, focusing on key issues and grouping them into themes [7]. Initially, we will use attribute (grammatical) and descriptive (elemental) first-cycle coding techniques. We are choosing these coding techniques for four reasons. First, attribute coding will enable us to separate participant demographic information into manager, non-manager, white collar and blue collar workers for future management and reference [18]. Second, descriptive coding will enable us to break down interview data [16] so that we can summarize topics into words and phrases (units) that can be interpreted [18]. Third, units from descriptive coding will serve as foundation for second or third cycle coding if needed [18]. Lastly, breaking data down into words and phrases will provide us with more understanding of the data and help us see evidence through multiple perspectives without being biased towards participant impressions and behaviors [16].

After coding, we will perform 'within' case analyses for each case by performing the following tasks. First, we will look for patterns among codes and group similar codes into sub-categories without regard to descriptive attributes (first-level groupings). Next, we will repeat this process for the sub-categories and further group similar sub-categories into categories (second-level groupings). Lastly, we will repeat the process for categories and group similar categories into themes (third-level groupings). We will perform these tasks such that we arrive at as many codes, sub-categories, categories and themes for each case for each of the research questions as recommended by Saldaňa [18]. This will find commonalities across managers and non-managers. Then, we will repeat the process within the descriptive categories to determine differences between managers, long-time employees, genders (if there is enough variation), age groups, and so on to the extent possible.

After within-case analyses, we will perform cross-case analyses. This analysis will be conducted in a similar manner. First, without regard to descriptive attributes, we will analyze similarities of themes across the groups. Then, we will analyze mindful of the descriptive categories to determine the extent to which the different grouping alter the feelings, attitudes, and perspectives of the respondents. From the analyses, we will seek to define the core category around which the basis for theoretical foundations can be developed [18].

3.5   Case Interpretation

After analysis, we will interpret our findings and draw conclusions based on our research questions [7]. This will be done separately for each of the themes that we will be developing from 'within' and cross case analyses for each research question. We plan to follow Toulmin's [19] argument model, to develop claims, provide evidence of those claims, and warrants to develop logic about the claims, based on our academic and past research credentials. The steps to analyzing the observational data come first, to note the event. The prior analysis of within and between cases analysis will surface events, pattern categories and, relationships between them. The categories become the claims, and the events become the evidence of those claims. The warrants are anecdotes about the circumstance of the stories told by the respondents and any pre or post comments that might relate. Evidence both within and between categories of similarities and differences will be used to develop theoretical claims.

For research question 1, we will discuss how themes generated from the interview data explain how SPMS provide job-related information. Role ambiguity and how it is affected by the presence, use, and feedback in the SPMS will be discussed at all levels and used to provide the validating warrants of claims relating to similarities and differences between the groups.

The process will be repeated for research question 2, discussing how role ambiguity, if present, relates to individual performance and, if the SPMS helps or hinders the perceptions of role ambiguity and how.

4 Conclusions

In this research, we propose that the SPMS—employee performance phenomenon needs to be further explored for a better understanding. We reviewed the literature and proposed a multiple qualitative case study research with a subjective constructionist worldview as appropriate for exploring the phenomenon. This research is important because prior research that has studied the phenomenon have mostly been quantitative studies with little focus on qualitative data. In addition, the phenomena have been studied with positivist worldviews with little attention to constructionist worldviews. Most importantly, the phenomenon by which SPMSs affect performance is not well understood, as there are conflicting findings in the literature. To resolve the conflicts, research questions on how SPMS affect role ambiguity and how role ambiguity affects performance are proposed to guide this study. In exploring the SPMS—performance phenomenon, the research questions will be the guide for collecting primary (interviews) and secondary data (documents on SPMS, employee performance evaluations and organization mission, vision, and strategy). Data will be collected from five business segments—Wholesale, retail, generation, mining, and transmission, of a utility company located in the Southern United States. We selected this organization because each of the five business segments has distinct management teams and organizational cultures that will help make our findings more robust. After exploring the phenomenon, we expect our findings, as outcomes, to (a) bridge the gap in the literature and (b) provide a further understanding of how SPMS affects individual performance in organizations. This will keep organizations well informed about individual outcomes that may result from implementing SPMS.

Appendix

The Interview Protocol

1. Tell me about how you spend your day at work. What are the major elements of your job?

2. Are there any parts of your job you are not clear about how to perform or not clear about what is expected of you? [review job as described above]
3. What information do you need from outside your job to do your work? Where does it come from? Do you always have it? Do you ever think you need more or other information to do your job? What type? Where might is come from? Why don't you have it?
4. In your opinion, what makes you a successful employee of this organization? Are there some elements of your job performance that might be improved? What and how?
5. Do you pay attention to the firm's SPMS and its feedback? How do you use it? What impact does it have on your job?
6. In a few words, can you describe the vision, mission, and strategy of your organization?
7. How does your organization's business model work relative to your job?
8. From the vision, mission and strategy you describe, describe your role in their implementation?
9. In your opinion, how does your role help your organization achieve its strategy?
10. A strategic performance management system, also known as SPMS, is used in your organization. Does it apply to your job? And how? What are benefits of SPMS to you? To your job?
11. To what extent and how does the SPMS help you do your job? To what extent and how does the SPMS help you understand your job duties?
12. What kind of information does your SPMS provide you? How is this information relevant to your job?
13. What do you do when you have questions about your job?
14. On what criteria are your performance evaluations based?
15. From your experience, how does the SPMS affect your performance evaluation?
16. Do you think your performance evaluation might be different if you had a better understanding of your job duties? Could the SPMS provide that understanding?
17. Does your performance evaluation process refer to the SPMS or your SPMS performance?

References

1. Asare, E., Conger, S.: Strategic performance measurement systems and employee performance. In: Rocha, A., Reis Paulo, L., Cota Perez, M., Suarez, S.O., Goncalves, R. (eds.) 11th Iberian Conference on Information Systems and Technologies, pp. 88–92. AISTI, Gran Canaria, Canary Islands, Spain (2016)
2. Bisbe, J., Malagueño, R.: Using strategic performance measurement systems for strategy formulation: does it work in dynamic environments? Manag. Account. Res. 23(4), 296–311 (2012). doi:10.1016/j.mar.2012.05.002

3. Burney, L., Henle, C., Widener, S.: A path model examining the relations among strategic performance measurement system characteristics, organizational justice, and extra- and in-role performance. Acc. Organ. Soc. **34**(3–4), 305–321 (2009). doi:10.1016/j.aos.2008.11.002
4. Burney, L., Widener, S.K.: Strategic performance measurement systems, job-relevant information, and managerial behavioral responses—role stress and performance. Behav. Res. Account. **19**, 43–69. http://search.proquest.com/docview/203298201?accountid=7106 (2007)
5. Burney, L., Widener, S.K.: Behavioral work outcomes of a strategic performance measurement system-based incentive plan. Behav. Res. Account. **25**, 115–143. http://search.proquest.com/docview/1449191526?accountid=7106 (2013)
6. Choi, J., Hecht, G.W., Tayler, W.B.: Lost in translation: the effects of incentive compensation on strategy surrogation. Account. Rev. **87**, 1135–1163. doi:10.2139/ssrn.1464803 (2012)
7. Creswell, W.J.: Qualitative Inquiry and Research Design, 3rd edn. SAGE Publications Ltd., Thousand Oaks, California (2013)
8. Eatough, E.M., Chu-Hsiang, C., Miloslavic, S.A., Johnson, R.E.: Relationships of role stressors with organizational citizenship behavior: a meta-analysis. J. Appl. Psychol. **96**(3), 619–632 (2011). doi:10.1037/a0021887
9. Hall, M.: The effect of comprehensive performance measurement systems on role clarity, psychological empowerment and managerial performance. Acc. Organ. Soc. **33**(2–3), 141–163 (2008). doi:10.1016/j.aos.2007.02.004
10. Kaplan, R.S., Norton, D.P.: The Balance Scorecard: Translating Strategy into Action. Harvard Business School Press, Boston, MA (1996)
11. Kaplan, R.S., Norton, D.P.: The Strategy-Focused Organization: How Balance Scorecard Companies Thrive in the New Business Environment. Harvard Business School Press, Boston, MA (2001)
12. Leach-Lopez, M.A., Stammerjohan, W.W., McNair, F.M.: Differences in the role of job-relevant information in the budget participation-performance relationship among U.S and Mexican managers: A question of culture or communication. J. Manag. Account. Res. **19**, 105–136 (2007)
13. Mintzberg, H., Ahlstrand, B., Lampel, J.: Strategy Safari: A Guided Tour Through the Wilds of Strategic Management. Free Press, NY, New York (1998)
14. Onyemah, V.: Role Ambiguity, Role Conflict, and Performance: Empirical Evidence of an Inverted-U Relationship. J. Pers. Sell. Sales Manag. **28**(3), 299–314 (2008). doi:10.2753/PSS0885-3134280306
15. Ordanini, A., Miceli, L., Pizzetti, M., Parasuraman, A.: Crowd-funding: transforming customers into investors through innovative service platforms. J. Serv. Manag. **22**, 443–470 (2011)
16. Parisi, C.: The impact of organisational alignment on the effectiveness of firms' sustainability strategic performance measurement systems: an empirical analysis. J. Manag. Gov. **17**(1), 71–97 (2013). doi:10.1007/s10997-012-9219-4
17. Rapiah, M., Hui, W.S., Ibrahim, R., Aziz, A.: The relationship between strategic performance measurement systems and organisational competitive advantage. Asian-Pac. Manag. Account. J. **5**(1), 1–20. http://9icsts2014.um.edu.my/filebank/published_article/6431/AJBA_5.pdf (2014)
18. Saldaña, J.: The Coding Manual for Qualitative Researchers, 2nd edn. SAGE Publications Ltd., Thousand Oaks, California (2013)
19. Toulmin, S.: The Uses of Argument, 15th edn. Cambridge University Press, London (2003)
20. Yin, R.K.: Case study research: design and methods. In: Bickman, L., Rog, D.J. (eds.) Essential Guide to Qualitative Methods in Organizational Research, vol. 5. Sage Publications. doi:10.1097/FCH.0b013e31822dda9e (2009)

Advanced Radial Approach to Resource Location Problems

Marek Kvet

Abstract This paper deals with the optimal resource location problems used for emergency service system designing. Due to limited budget or other technological restrictions, particular mathematical models usually take the form of the weighted p-median problem, in which the numbers of served users and possible service center locations, from which the associated service is provided, take the value of several hundreds or thousands. A standard objective in such formulation assumes minimization of total disutility, like social costs. The social costs are often proportional to the distance travelled by all system users to the nearest located source of provided service. As the access of population to the service is performed by transportation means operating on the underlying transportation network, the design of suitable deployment of the service centers belongs to hard combinatorial problems. This contribution deals with an approximate approach based on radial formulation with homogenous system of radii. Presented approach enables to solve large instances in admissible time making use of a universal IP-solver. Special attention is paid to possible adjustment of the approximate approach to the exact method based on useful features of the radial formulation.

1 Introduction

Optimization and methods of operational research play a very important role in many areas of human life. Results of particular research can be applied in health-care systems [3, 10], bioengineering [13, 14], computer science and automation [15], public sector and service systems designing [5, 16], traffic solutions and in many other fields connected with discrete optimization [2]. Within this chapter, we deal with effective approximate solving method suggested for large instances of these discrete network location problems and focus on the emergency service system designing. Making

M. Kvet (✉)
Faculty of Management Science and Informatics,
University of Žilina, Univerzitná 8215/1, 010 26 Žilina, Slovak Republic
e-mail: marek.kvet@fri.uniza.sk

© Springer International Publishing AG 2018
Á. Rocha and L.P. Reis (eds.), *Developments and Advances in Intelligent Systems and Applications*, Studies in Computational Intelligence 718,
DOI 10.1007/978-3-319-58965-7_3

29

use of presented approach we can obtain either the optimal or a good near optimal solution of the problem in a short time.

The emergency service system designing is a challenging task for both system designer and operational researcher. As the first one searches for a tool, which enables to obtain service center deployment satisfying future demands of the system users, the second one faces the necessity of completing the associated solving tool. Emergency service system efficiency is considerably influenced by deployment of the service centers, which send emergency vehicles to satisfy demands on service at system users' locations. The number of service providing centers must be limited due to economic and technological restrictions regardless the case whether the service is delivered to users or the users travel for service. Thus the emergency service system design consists in locating a limited number of service centers at positions from a given finite set to optimize quality characteristic of the designed system from the point of the served system users. As concerns the characteristic of quality, various objective functions have been formulated. The most frequently used quality criterion of the design takes into account some evaluation of the users' discomfort, i.e. social costs. The social costs are often proportional to the distances from served users to the nearest source of provided service. These costs can be denoted as disutility, to which the system user is exposed. In other words, the standard approach to the emergency service system design assumes that each user is served from the nearest located service center or from the center, which offers the smallest individual disutility [10]. We also assume, that each service center has enough capacity to serve all assigned users. If the quality characteristic corresponds to service accessibility of an average user, then the emergency service system design can be tackled as a weighted p-median problem, which was studied in [4, 6, 10, 16] and others.

If a large instance of the problem is described by a location-allocation model, then the model size often exceeds any acceptable limit for available optimization software equipped with the branch and bound method. It must be realized, that the numbers of served users and possible service center locations, from which the associated service is provided, may take the value of several hundreds or thousands [1]. Another way consists in development of specific software tool for particular emergency service system design, but this way is both costly and long term demanding. To avoid this obstacle, the approximate approach based on radial formulation has been developed [4, 6]. On the contrary to the ZEBRA algorithm introduced in [4], our approximate approach is based on homogenous system of radii given by so-called dividing points [6, 7].

The keystone of suggested approach developed primary for solving the standard p-median problem and reported in [6, 7] consists in minimization of the upper bound of the original objective function. As the radial formulation of the problem avoids assigning a center to a user location like it is common in the location-allocation approach, the radial model is smaller than the location-allocation one. In addition, the solving method used in the IP-solvers converges much faster. On the contrary to common heuristics, this approach enables determination of a lower bound of the optimal objective function value and thus it is possible to evaluate the accuracy of the obtained result. If necessary, presented radial formulation enables to make a trade-off between a little loss of optimality and the computational time of large instance solving process [12]. As concerns the quality of resulting solution, it must be noted, that its accuracy depends on suitable determination of so-called dividing points, which define homogenous set of radii. Thus, the way of disutility range partitioning plays a very important role in the approximate approach [7–9]. In this paper, we present both exact and approximate approaches to the weighted p-median problem and study their limits, solution accuracy and computational time demands. Special attention is paid to a two-phase algorithm, which enables to obtain the optimal solution of large problem instances using the radial formulation with a reduced set of possible disutility values. To reduce the disutility range, an advanced min-max model is solved [11]. The main goal of this study is to present a useful tool, that can be easily implemented within a commercial IP-solver. We also give an overview of numerical results to prove the usefulness of suggested technique. To obtain both exact and approximate solution of tested benchmarks, the optimization software Xpress-IVE was used.

2   Mathematical Model Formulation

The emergency service system design problem with minimal total users' disutility is often formulated as a task of location of at most p service centers (p is a positive integer value) from a given set so that the sum of individual disutility values of each user coming only from the nearest located service center is minimal. Thus, the standard approach takes into consideration the average system user.

To describe the above-mentioned problem by means of mathematical programming, let the symbol I denote the set of possible service center locations (service providers). Here, it is assumed that all service centers have equal setup cost and enough capacity to serve all users. Furthermore, let J be used to define the set of possible users' locations (service recipients). Each user location is represented by a specific point. The symbol b_j denotes the number of users sharing the location $j \in J$. The disutility for a user at the location j following from the possible center location i is denoted as non-negative d_{ij}. This problem is also known as the weighted p-median problem, which is broadly discussed in [1, 3, 7, 10, 12] from the viewpoint of solving techniques suggested for huge instances. Under presented assumptions, we can state the problem as follows.

$$\text{Minimize} \quad \left\{ \sum_{j \in J} b_j min \left\{ d_{ij} : i \in I_1 \right\} : I_1 \subseteq I, |I_1| \leq p \right\} \quad (1)$$

The symbol I_1 denotes a subset of the set I of all possible service center locations. To formulate the location-allocation model of this problem with linear objective function, the following decisions must be modeled by variables introduced below.

The basic decisions in any solving process of the weighted p-median problem concern the location of service centers at the network nodes from the set I so that the

sum of users disutility contributions is minimal and the number of located centers
does not exceed the value of p. To model this decision at particular location, we
introduce a zero-one variable $y_i \in \{0,1\}$, which takes the value of 1, if a center
should be located at the location i, and it takes the value of 0 otherwise. In addition,
the allocation variables $z_{ij} \in \{0,1\}$ for each $i \in I$ and $j \in J$ are introduced to assign
a user location j to a possible service center location i by the value of 1. To meet the
problem requirements, the decision variables y_i and also the allocation variables z_{ij}
have to satisfy the following constraints. Thus, the location-allocation model can be
formulated according to [7] by (2)–(7).

$$\text{Minimize} \qquad \sum_{j \in J} b_j \sum_{i \in I} d_{ij} z_{ij} \qquad\qquad (2)$$

$$\text{Subject to:} \qquad \sum_{i \in I} z_{ij} = 1 \qquad for \quad j \in J \qquad\qquad (3)$$

$$\sum_{i \in I} y_i \leq p \qquad\qquad (4)$$

$$z_{ij} \leq y_i \qquad for \quad i \in I, \quad j \in J \qquad\qquad (5)$$

$$y_i \in \{0,1\} \qquad for \quad i \in I \qquad\qquad (6)$$

$$z_{ij} \in \{0,1\} \qquad for \quad i \in I, \quad j \in J \qquad\qquad (7)$$

In the above model, the objective function (2) minimizes the sum of disutility
values between the system users and the nearest located service centers. The allocation constraints (3) ensure that each system user is assigned to exactly one possible
service center location. Link-up constraints (5) enable to assign a user location j to
a possible center location i only if the service center is located at this location and
the constraint (4) bounds the number of located service centers by p. The problem
described by terms (2)–(7) can be rewritten to a form acceptable by a modeler of
integrated optimization environment and solved by the associated IP-solver.

The optimal solution searching process for a large problem instance described by
the location-allocation model (2)–(7) requires huge memory capacity and computational time. That is the main reason, why algorithms integrated into common decision support tools often fail when solving such problems [7]. Therefore the location-
allocation formulation does not hold for practical usage. On the other hand, it is obvious from the constraints (3), that there is only one variable z_{ij} for each user location
$j \in J$, which takes the value of 1. It means that only one distance from each column
of the matrix $\{d_{ij}\}$ becomes relevant and this value can be approximated without the
necessity of allocation variables z_{ij}. This observation leads to a radial model, which
takes the advantages of the set-covering problem as easy implementation and good
solvability.

3 Radial Approach to Emergency Service System Design

3.1 Exact Model

The necessity of solving large instances of the weighted p-median problem has led to many approximate approaches based on heuristic and metaheuristic algorithms. Within this paper, we deal with a radial formulation of the problem, which avoids assigning the individual users' location to some of located service centers and deals only with the information, whether some service center is or is not located in given radius from the users' location. In other words, information about the number of service centers located in given radius from the users' location is used instead of formalized knowledge of the nearest located service center.

Mentioned radial formulation of the p-median problem has appeared in two different versions. One of them [4] creates a unique system of zones for each users' location and an individual radius corresponds with concrete disutility between the users' location and some possible service center. Our approach is based on the second version [6, 7], where the range of all considered disutility values is partitioned by so-called dividing points and an individual radius corresponds with the position of a dividing point. In this version, the same system of radii is applied to each users' location. It follows from the fact that there is only a finite number of various disutility values in the matrix $\{d_{ij}\}$, which can enter the optimal solution of the associated weighted p-median problem. Note that none of the largest, second largest and so on up to $p-1$ largest distances from given users' location j to the set of all possible service center locations can be contained in any optimal solution. Let the mentioned set of $m+1$ different disutility values form an increasing sequence $d_0 < d_1 < \cdots < d_m$. Without any loss of generality, we can assume that d_0 is equal to zero; in the opposite case, we can reduce each item of the matrix subtracting the minimal value.

The radial formulation is based on the construction of constraint (8) for each users' location j. As above, the variable $y_i \in \{0,1\}$ models the decision of service center location at $i \in I$ by the value of 1. This notation is used also in the remainder of this paper. In addition, an auxiliary variable x_{js} is used to indicate by the value of one that no service center is located in the radius d_s from the users' location j. Otherwise; the variable takes the value of one. To complete the constraint, the zero-one constants a_{ij}^s must be defined for the disutility d_s and the users' location j and for each $i \in I$ so that the constant a_{ij}^s is equal to 1, if the disutility d_{ij} perceived by the users located at j from the possible center location i is less than or equal to d_s, otherwise a_{ij}^s is equal to 0.

$$x_{js} + \sum_{i \in I} a_{ij}^s y_i \geq 1 \tag{8}$$

The constraints (8) ensure that the variable x_{js} is allowed to take the value 0, if there is at least one service center located in the radius d_s from given users' location j. In such case, the second term on the left-hand-side of (8) takes the value corresponding to the number of located service centers in the radius d_s and if this number is greater than or equal to one, then x_{js} can take the value of zero unless the constraint is broken. When the value of variable is pushed down by an optimization process, then it takes the value of one, if the second term takes the value of zero and otherwise it takes the value of one.

If the objective of emergency service system design problem consists in minimization of disutility perceived by an average user from the nearest located service center, then the associated model takes the form of the weighted p-median problem, and its exact radial formulation is as follows. The variables x_{js} and constants a_{ij}^s are defined for each users' location $j \in J$ and $s \in [0..m-1]$ and also the symbols e_s are introduced to denote the differences $d_{s+1} - d_s$ for $s \in [0..m-1]$.

$$\text{Minimize} \qquad \sum_{j \in J} b_j \sum_{s=0}^{m-1} e_s x_{js} \qquad (9)$$

$$\text{Subject to:} \qquad x_{js} + \sum_{i \in I} a_{ij}^s y_i \geq 1 \qquad for \quad j \in J \quad and \quad s = 0, 1 \ldots m-1 \quad (10)$$

$$\sum_{i \in I} y_i \leq p \qquad (11)$$

$$x_{js} \geq 0 \qquad for \quad j \in J \quad and \quad s = 0, 1 \ldots m-1 \qquad (12)$$

$$y_i \in \{0, 1\} \qquad for \quad i \in I \qquad (13)$$

In this model, the constraints (10) ensure that the variables x_{js} are allowed to take the value 0, if there is at least one service center located in the radius d_s from the users' location j. As the minimization process applied on (9) pushes all included variables x_{js} down to the zero value and each of these variables is limited from below either by the value of one or by the value of zero according to (8), the variable x_{js} can get only one of these values unless obligatory 0–1 constraints must be included into the model. The constraint (11) puts the limit p on the number of located service centers.

Even if the problem in the radial form is smaller and easier to solve than instances described by the location-allocation formulation, memory insufficiency may occur, when large instances of the weighted p-median problem are solved using commercial IP-solver. These instances are characterized by high value m in the sequence $d_0 < d_1 < \cdots < d_m$. In such case, the presented exact approach can be converted into the approximate one, which pays by controlled loss of accuracy for lower memory demand and smaller computational time.

3.2 Approximate Formulation with Dividing Points

The biggest disadvantage of the exact radial model (9)–(13) consists in the number of elements included in the sequence $d_0 < d_1 < \cdots < d_m$ mainly when large problem instances are solved. This fact makes this approach disputable from the viewpoint of practical usage. In the case, when the value of m is too high to process the complete sequence $d_0 < d_1 < \cdots < d_m$, there is a way to overcome the difficulties brought by big size of the problem (9)–(13). This method makes use of the concept of so-called dividing points introduced in [6, 7] and converts the approach to approximate.

Here, the radial formulation is based on the idea of upper or lower approximation of the individual user's disutility. To obtain the upper approximation of the original objective function value, the range $[d_0, d_m]$ of all possible $m + 1$ disutility values is partitioned into $v + 1$ zones. The zones are separated by a finite ascending sequence of dividing points $D_1, D_2, \ldots, D_v$ chosen from the values $d_1 < d_2 < \cdots < d_{m-1}$. Let us define also $D_0 = d_0$ and $D_{v+1} = d_m$. Then the zone s corresponds with the semi-closed interval $(D_s, D_{s+1}]$ for $s = 0, \ldots, v$. The length of the s-th interval is denoted by e_s. The auxiliary variable x_{js} for $s = 0, \ldots, v$ is here not related to the disutility d_s like in the exact model (9)–(13), but it takes the value of 1, if the disutility perceived by the users located at $j \in J$ from the nearest located service center is greater than the value of dividing point D_s. Otherwise, x_{js} takes the value of 0. Then the expression $e_0 x_{j0} + e_1 x_{j1} + \cdots + e_v x_{jv}$ constitutes the upper approximation of the disutility d_{j*} perceived by the users located at j from the nearest located center. If the disutility d_{j*} belongs to the interval $(D_s, D_{s+1}]$, then the value of D_{s+1} is the upper estimation of d_{j*} as it is shown in the Fig. 1.

Similar changes have to be performed with coefficients a_{ij}^s. In the approximate model, this constant is equal to 1, if the disutility d_{ij} between the users' location j and the possible service center location i is less than or equal to the value of dividing point D_s, otherwise a_{ij}^s is equal to 0. After these preliminaries, the radial model (9)–(13) can be solved to get the approximate solution of the associated weighted p-median problem. To obtain a lower bound of the original problem optimal solution, we can use the above introduced dividing points and the associated zone widths, and express

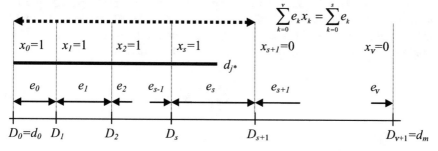

Fig. 1 *Upper* approximation of d_{j*} using zone widths e_s and auxiliary variables x_{js}. The *upper* approximation of d_{j*} is denoted by *thick dotted line* at the *top* of figure

the lower bound of d_{j*} as $e_0 x_{j1} + e_1 x_{j2} + \cdots + e_{v-1} x_{jv}$. The lower approximation of d_{j*} enables to evaluate the quality of the resulting design and measure the maximal deviation from the unknown optimal solution.

Since the approximate radial approach to the weighted p-median problem provides only lower or upper approximation of the original objective function (2), the corresponding real objective function value $ObjF(y)$ must be computed according to (14) based on the vector y of location variables y_i for $i \in I$.

$$ObjF(y) = \sum_{j \in J} b_j min \left\{ d_{ij} : i \in I, y_i = 1 \right\} \tag{14}$$

The keystone of the approximate approach consists in selection of the dividing points, which have direct influence on the result accuracy and computational time. The number v of dividing points influences the size of the radial model (9)–(13) as concerns either the number of variables x_{js}, which are introduced for each user location $j \in J$ and each zone s, or the number of the constraints (10). Obviously, only limited number v of the dividing points $D_1, D_2, \ldots, D_v$ can be used to keep the size of the radial model (9)–(13) in a solvable extent. This restriction impacts the deviation of the approximate solution from the exact one. On the other hand, the smaller the number of dividing points is, the bigger inaccuracy afflicts the approximate solution [7, 12]. The v dividing points can be chosen only from the set of values $d_0 < d_1 < \cdots < d_m$, where $D_0 = d_0$ and $D_{v+1} = d_m$. Let each value d_h for $h = 0, 1, \ldots, m$ have a frequency N_h of its occurrence in the matrix $\{d_{ij}\}$. In the suggested approach, we start from the hypothesis that the disutility value d_h occurs in the resulting solution n_h times and that is why the deviation of this value from its approximation encumbers the total deviation proportionally to n_h. The disutility d_{j*} perceived by the users located at j from the nearest located center can be only estimated taking into account that it belongs to the interval $(D_s, D_{s+1}]$. The maximal deviation of the upper estimation D_{s+1} from the exact value d_{j*} is $D_{s+1} - D_s$. If we were able to anticipate the frequency n_h of each d_h in the unknown optimal solution, we could minimize the total deviation of the upper approximation from the optimal solution by solving the problem (15)–(19) to obtain convenient deployment of dividing points. Let us introduce zero-one variables u_{ht} for each possible position t of the dividing point $d_t (t = 1 \ldots m)$ and for each possible position h of the preceding value $d_h (h = 0 \ldots t)$. If the disutility value d_h belongs to the interval ending by the dividing point d_t, then the variable u_{ht} takes the value of 1, otherwise u_{ht} equals 0. If u_{tt} is equal to 1, then the disutility d_t corresponds with the dividing point.

$$\text{Minimize} \quad \sum_{t=1}^{m} \sum_{h=1}^{t} (d_t - d_h) n_h u_{ht} \tag{15}$$

$$\text{Subject to:} \quad u_{(h-1)t} \leq u_{ht} \quad for \quad t = 2, \ldots, m \quad and \quad h = 2, \ldots, t \tag{16}$$

$$\sum_{t=h}^{m} u_{ht} = 1 \qquad for \quad h = 1, \ldots, m \tag{17}$$

$$\sum_{t=1}^{m-1} u_{tt} = v \tag{18}$$

$$u_{ht} \in \{0, 1\} \qquad for \quad t = 1, \ldots, m \quad and \quad h = 1, \ldots, t \tag{19}$$

The link-up constraints (16) ensure that the disutility value d_{h-1} belongs to the interval ending with d_t only if each other disutility between d_{h-1} and d_t belongs to this interval. Constraints (17) assure that each disutility value d_h belongs to some interval and the constraint (18) enables only v dividing points to be chosen. After the problem (15)–(19) is solved, the nonzero values of u_{tt} indicate the disutility values d_t corresponding with dividing points. The associated approximate solving technique for the weighted p-median problem consists of the estimation of the relevance n_h for $h = 0, 1, \ldots, m$, solving the dividing points deployment problem (15)–(19) and then solving the radial model (9)–(13).

The above approximate approach is obviously based on the "relevance" of any disutility d_h, which expresses the strength of our expectation that the disutility value d_h will be a part of the unknown optimal solution, which is searched for. We have developed and explored several ways of the relevance estimation [7, 9]. All suggested approaches to the disutility relevance estimation have been compared from the viewpoint of solution accuracy for given constant number v of dividing points. Based on our previous research aimed at proper estimation of n_h, the most suitable proved to be so-called "shifted exponential approach" [8], which estimates the relation between n_h and N_h by formulae (20)–(22).

$$n_h = N_h g(h) \tag{20}$$

The function $g(h)$ is equal to 1 for each $h \le h_{\text{crit}}$ and it is defined by (21) for $h > h_{\text{crit}}$.

$$g(h) = e^{h_{\text{crit}} - h} \tag{21}$$

The constant h_{crit} is a parameter of the approach called "critical value", which can be determined according to the expression (22).

$$h_{\text{crit}} = min \left\{ h \in Z^+ : \sum_{u=0}^{h} N_u \ge \frac{\sqrt{|I|/p}}{p} \sum_{t=0}^{m} N_t \right\} \tag{22}$$

3.3   Reduction of the Disutility Range for the Exact Model

Despite the fact that presented radial approach was originally designed as effective heuristic modelling technique providing enough accurate solution of large instances in a short time [6, 7, 12], the necessity of obtaining the optimal solution has led to several adjustments of the radial approach, which enable to comply with the model complexity and overcome the weakness of previously discussed location-allocation formulation. When constructing the radial model (9)–(13), two contradictory demands are put on constitution. The first demand consists in maximal accuracy of the obtained solution. This demand may issue into very large size of associated model caused by big number of elements included in the sequence $d_0 < d_1 < \cdots < d_m$. This fact impacts also the computational time and memory requirements. The second demand asks for acceptable computational time and tries to minimize the model size as much as possible. Therefore, we present here another concept, which enables to comply with both demands to certain extent. The idea is based on a two-phase algorithm. The first step reduces the disutility sequence $d_0 < d_1 < \cdots < d_m$ for the second phase, in which the radial model (9)–(13) is solved exactly, so the optimal solution of the original problem can be obtained without the dividing points computation. Mentioned reduction of the sequence $d_0 < d_1 < \cdots < d_m$ follows from simple assumption: There is pretty high probability that higher disutility values will not be included in the optimal solution and that is why we try to find the maximal relevant disutility less than d_m. This value can be estimated by proper increasing the result of "min-max" problem, which minimizes the disutility perceived by the worst situated users. It must be noted, that the disutility perceived by the worst situated users in the original weighted p-median problem is usually higher than the result of the "min-max" problem. The original problem minimizes the disutility of average users and the position of the worst situated users is not taken into account, while the "min-max" problem minimizes the disutility perceived by the worst situated users.

The "min-max" emergency service system design problem can be described by the above-mentioned denotations. As above, the variables y_i model the decision of service center location at particular location $i \in I$ and the auxiliary variables x_{js} are used to indicate whether the disutility perceived by users located at $j \in J$ is greater than the value d_s. Particular radial model of this problem can be formulated as follows.

$$\text{Minimize} \quad h \tag{23}$$

$$\text{Subject to:} \quad x_{js} + \sum_{i \in I} a_{ij}^s y_i \geq 1 \quad for \quad j \in J \quad and \quad s = 0, 1 \ldots m - 1 \tag{24}$$

$$\sum_{i \in I} y_i \leq p \tag{25}$$

$$\sum_{s=0}^{m-1} e_s x_{js} \leq h \quad for \quad j \in J \tag{26}$$

$$x_{js} \geq 0 \qquad for \quad j \in J \quad and \quad s = 0, 1 \ldots m - 1 \qquad (27)$$

$$y_i \in \{0, 1\} \qquad for \quad i \in I \qquad (28)$$

$$h \geq 0 \qquad (29)$$

The objective function (23) represented by single variable h gives the upper bound of all perceived disutility values. The constraints (24) ensure that the variables x_{js} are allowed to take the value 0, if there is at least one service center located in the radius d_s from the users' location j. The constraint (25) puts the limit p on the number of located service centers. Finally, the constraints (26) ensure that each perceived disutility is less than or equal to the upper bound h.

As it has been already mentioned, the sequence $d_0 < d_1 < \cdots < d_m$ may contain too many elements and that is why the model (23)–(29) is not suitable for large instances. Furthermore, the objective function (23) together with the link-up constraints (26) make the model hard to solve and ruin the useful features of the radial formulation. So the solving process performs really bad. On the other hand, presented "min-max" model can be notably simplified to avoid the link-up constraints (26) and the necessity of processing the whole disutility range $[d_0, d_m]$ in one turn. Possible modification follows from the structure of the variables x_{js} introduced for each user location j and each index $s \in [0 .. m - 1]$. Consider the explanation of the radial approach in Sect. 3.1 and Fig. 1 in Sect. 3.2. As it is reported also in [4], for any optimal solution vector x, given user location $j \in J$, subvector $x_j = (x_{j0}, x_{j1}, \ldots, x_{j,m-1})$ has the following structure: $x_j = (1, 1, \ldots, 1, 0, 0, \ldots, 0)$. That is, it begins with the value of one in all its components until it reaches some index, where the first zero appears. From there, all the components have the value of 0. This property holds for every users' location j, and it is an immediate consequence of the hierarchical structure of the constraints because of (30).

$$\{i : d_{ij} \leq d_s\} \subseteq \{i : d_{ij} \leq d_{s+1}\} \qquad (30)$$

This implies that obtaining the optimal solution of the "min-max" problem consists in finding the smallest disutility d_{s*}, for which all variables x_{js*} take the value of 0. Then the value of d_{s*} represents the maximal relevant disutility for the worst situated users—the result of the "min-max" problem (23)–(29). The subscript $s*$ can be searched by the bisection method. To summarize suggested method, the bisection radial approach makes use of the radial model, but it uses only its reduced form in determining whether there is a solution with the objective function value less than or equal to the given disutility value d_s. In this model, the zero-one variables $y_i \in \{0, 1\}$ for $i \in I$ are also used as before. The variables x_j are introduced to indicate whether the disutility of the users at the location $j \in J$ following from the nearest located center is greater than d_s. In such case, the variable takes the value of 1. The constants a_{ij} for each $i \in I$ and $j \in J$ need to be set in each iteration for particular value of d_s. The coefficient a_{ij} is equal to 1, if the disutility d_{ij} is less than or equal to d_s, otherwise a_{ij} is equal to 0. The corresponding model for one iteration of the process is formulated as follows.

$$\text{Minimize} \qquad \sum_{j \in J} x_j \tag{31}$$

$$\text{Subject to:} \qquad x_j + \sum_{i \in I} a_{ij} y_i \geq 1 \qquad for \quad j \in J \tag{32}$$

$$\sum_{i \in I} y_i \leq p \tag{33}$$

$$x_j \geq 0 \qquad for \quad j \in J \tag{34}$$

$$y_i \in \{0, 1\} \qquad for \quad i \in I \tag{35}$$

In this simple covering model, the objective function (31) represents the number of user locations, where the perceived disutility is greater than d_s. The constraints (32) ensure that variables x_j are allowed to take the value of 0, if there is at least one centre located within radius d_s extending from the user location j and constraint (33) limits the number of located service centres by p. With the reduced form of the radial approach (31)–(35), the dividing points are not needed, because the maximal relevant disutility d_{s^*} is searched by the bisection method applied on the whole range $[0, m-1]$. The number of iterations necessary for obtaining the value of s^* depends on the number of elements in the sequence $d_0 < d_1 < \cdots < d_m$.

Having solved the "min-max" problem, the resulting value d_{s^*} of the bisection process is used to reduce the sequence $d_0 < d_1 < \cdots < d_m$ for the second stage, in which the exact radial model (9)–(13) is solved to get the resulting emergency service system design. The sequence $d_0 < d_1 < \cdots < d_m$ is reduced in such a way that all its elements starting with d_0 up to the subscript m^*, where $s^* \leq m^* \leq m$, are taken. Suitable setting of m^* by increasing the value of s^* is hard and relative, because of these two assumptions:

1. If the value of m^* is too low, then we do not cover all relevant disutility values and there will be some variables x_{j,m^*-1} in the radial model (9)–(13), which take the value of 1. It means that the disutility perceived by those users is higher than d_{m^*}. Thus the resulting solution does not have to be optimal. In such case, we have to increase the value of m^* and solve the problem (9)–(13) again for more considered elements $d_0, d_1, \ldots, d_{m^*}$. If all variables x_{j,m^*-1} take the value of 0, all perceived disutility values are less than or equal to d_{m^*} and the resulting solution is optimal. The direct consequence of low value of m^* consists in high number of iterations, in which the radial model (9)–(13) has to be solved.

2. If the value of m^* is too high and the model size does not exceed solvable extent, then the optimal solution of the problem (9)–(13) is found immediately (all variables x_{j,m^*-1} take the value of 0), but the computer memory may be demanded uselessly, what is paid by higher computational time.

In our research it was found that the maximal disutility d_{max} occuring in the optimal solution of the weighted p-median problem described by model (9)–(13) mostly meets the following inequalities (36).

$$2d_{s^*} \leq d_{max} \leq 3d_{s^*} \tag{36}$$

Therefore we suggest to reduce the sequence $d_0 < d_1 < \cdots < d_m$ in such a way that $m^* = min\{2d_{s^*}, m\}$. In case, when the subscript m^* takes its value around 20-30, the radial model can be easily solved exactly. Otherwise, the exact model (9)–(13) keeps being time and memory demanding [12].

Finally, the exact method based on the radial formulation with reduced sequence of disutility values can be summarized into the following steps:

1. Use the bisection method and iteratively solve the radial "min-max" problem (31)–(35) to get the maximal disutility d_{s^*}.
2. Reduce the sequence of disutility values $d_0, d_1, \ldots, d_m$ so that $m^* = min\{2d_{s^*}, m\}$.
3. Solve the exact radial model (9)–(13) with the reduced set $d_0, d_1, \ldots, d_{m^*}$. Check the sum of the resulting values of all variables x_{j,m^*-1}. If the sum is greater than 0, increment the subscript m^* and repeat this step. Otherwise, the optimal solution of the original problem is obtained.

3.4 Summary

In Sect. 3, the radial approach to emergency service system design is reported. It can be used in three different versions. The basic radial formulation reported in Sect. 3.1 processes the whole disutility sequence $d_0, d_1, \ldots, d_m$ and thus enables to get the optimal solution of the original problem. If the size of the solved instance is too high, then the concept of dividing points may be applied and an approximate solution of the problem can be obtained in a short time. On the contrary to common heuristics, applying the lower approximation of the disutility values enables to evaluate the quality of the result measured by maximal deviation from the unknown optimal solution. This concept is reported in the Sect. 3.2. Finally, we studied the possibility of obtaining the optimal solution also in such cases, when the exact radial model fails. We have suggested a two-phase method, which takes the advantages of the set-covering problems and enables to solve large problem instances exactly. The proposed algorithm is described in Sect. 3.3.

4 Numerical Experiments

To compare the exact and approximate radial approaches to the emergency service system design with the location-allocation formulation, we performed the series of numerical experiments. The main goal of this Section is to provide the readers with

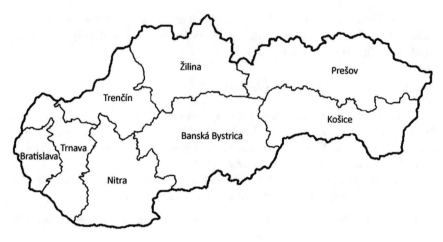

Fig. 2 Self-governing regions of Slovakia

achieved results and to summarize the findings, which confirm effectivity and flexibility of suggested modelling techniques.

All experiments were performed using the optimization software FICO Xpress 7.7 (64-bit, release 2014). The associated code was run on a PC equipped with the Intel Core i7 2630 QM processor with the parameters: 2.0 GHz and 8 GB RAM. Particular exact and approximate modelling techniques were tested on the pool of benchmarks obtained from the road network of Slovak Republic. The instances are organized so that they correspond to the administrative organization of Slovakia (Fig. 2).

The original problem comes from real emergency medical service system, where given number of ambulance stations should be deployed in the region to ensure the rescue service for the associated population. For each self-governing region, i.e. Bratislava (BA), Banská Bystrica (BB), Košice (KE), Nitra (NR), Prešov (PO), Trenčín (TN), Trnava (TT) and Žilina (ZA), all cities and villages with corresponding number of inhabitants b_j were taken. The constants b_j were rounded up to hundreds. The set of communities represents both the set J of users' locations and also the set I of possible center locations. It means that each community (even the smallest) may represent a possible service center location. Here we assume that each possible location of a service center has enough capacity to serve all users. The size of the set I for each self-governing region is shown in the following Table 1.

The network distance from a users' location to the nearest located service center was taken as an individual users' disutility. For each size of the set I, 11 different instances were solved. These instances differ in the value of parameter p, which limits the number of located service centers. The value of p was set in such a way, that the ratio of $|I|$ to p equals 2, 3, 4, 5, 10, 15, 20, 30, 40, 50 and 60 respectively. To enrich the set of benchmarks, three additional problems were solved. The emergency service system was designed also for the West Slovakia, which contains the regions of Bratislava, Banská Bystrica, Nitra, Trenčín, Trnava and Žilina. Here, the value of $|I|$ equals 1792 and p was set to 180. The second benchmark was created by joining the regions of Košice and Prešov as East Slovakia with $|I| = 1124$ and $p = 112$.

Table 1 Size of tested benchmarks for self-governing regions

Abbreviation	Self-governing region	Possible service center locations
BA	Bratislava	87
BB	Banská Bystrica	515
KE	Košice	460
NR	Nitra	350
PO	Prešov	664
TN	Trenčín	276
TT	Trnava	249
ZA	Žilina	315

Finally, the weighted p-median problem was solved also for the whole Slovakia, where the cardinality of the set I representing the possible service center locations takes the value 2916. Here, the value of parameter p was set to 273. This number corresponds to the real emergency medical service system in our country.

An individual experiment was organized so that each instance was solved exactly by the location-allocation approach described by the model (2)–(7). Then, the exact radial approach (9)–(13) was employed to compare the time complexity of both exact approaches. The following Table 2 contains the achieved results for selected small and middle-sized instances. The computational time in seconds is reported in columns denoted by *Time* and the value of *ObjF* corresponds to the objective function value of the optimal solution. Within the radial approach, we give also the cardinality $m + 1$ of the sequence $d_0, d_1, \ldots, d_m$.

It must be noted, that the number $m + 1$ of all possible disutility values included in the sequence $d_0, d_1, \ldots, d_m$ depends on the parameter p, which limits the number of located service centers. As we have shown in [7, 12], when minimizing the objective function value of the radial model, the $p - 1$ largest disutility values from each matrix column can be excluded as non-relevant. Thus, the value of $m + 1$ may differ for each benchmark even for the same self-governing region.

Despite the fact that we have solved only some small and middle-sized instances, the reported results have confirmed our presumptions. We can observe that the radial approach performs worse than the location-allocation one almost in all solved instances. Obtaining the exact solution by radial approach (9)–(13) requires processing of the whole sequence $d_0, d_1, \ldots, d_m$, that is dear-paid by high computational time. On the other hand, the location-allocation approach often fails when large instances are solved due to allocation variables and limited memory capacity. Therefore we have suggested and verified two other modelling techniques based on the radial formulation with homogenous system of radii.

The approximate radial approach with dividing points has been introduced in [6, 7] and its features and proper parameter settings have been explored in [8, 9, 12]. Within this paper, it is briefly studied mainly from the viewpoint of solution accuracy.

Table 2 Comparison of the location-allocation and radial exact approaches

Region	$|I|$	p	LocAlloc (2)–(7)		Radial (9)–(13)		
			Time [s]	ObjF	m + 1	Time [s]	ObjF
Žilina	315	158	5.17	2444	101	4.69	2444
	315	105	5.17	5594	115	6.30	5594
	315	79	5.91	8430	123	7.45	8430
	315	63	5.53	11125	128	8.03	11125
	315	32	5.33	20995	136	7.75	20995
	315	21	5.17	28548	139	7.44	28548
	315	16	5.34	34405	141	7.50	34405
	315	11	5.42	47092	143	7.39	47092
	315	8	5.63	58665	145	7.77	58665
	315	7	5.48	63865	148	7.83	63865
Prešov	664	332	43.27	2562	135	68.53	2562
	664	222	44.80	5518	157	90.14	5518
	664	166	47.64	8210	168	99.02	8210
	664	133	48.28	10604	180	95.56	10604
	664	67	61.11	20025	215	120.32	20025
	664	45	48.20	26981	223	106.47	26981
	664	34	51.45	32880	226	114.16	32880
	664	23	53.03	42409	229	98.86	42409
	664	17	54.24	51071	231	97.94	51071
	664	14	57.14	57815	233	95.19	57815

The exact solution (objective function value) resulting from the location-allocation model was taken as a referential, to which the solution obtained by the approximate radial approach was compared. Of course, it is not standard to compare the exact and approximate approaches, but it must be noted, that both methods use the same optimization environment equipped with the branch and bound method for integer programming problems. Furthermore, the exact solution is used here to evaluate the quality of the solution obtained by the approximate approach. Since the covering model provides only the approximation (upper bound) of the former objective function value, its real value must be computed according to (14). The accuracy of the solution can be generally evaluated by *gap* defined as follows: Let *ES* denote the objective function value of the exact solution computed by the location-allocation model (2)–(7) and let *CS* denote the objective function value of the approximate covering solution obtained by the expression (14) for the result of the model (9)–(13). Then the *gap* expresses the difference between these two values in percentage of the exact solution, what can be formulated by the following expression (37).

$$gap = \frac{|CS - ES|}{ES} * 100 \qquad (37)$$

Table 3 Results of numerical experiments with the approximate radial approach for the self-governing regions of Slovakia

Region	$|I|$	p	LocAlloc		Radial—approx		
			Time (s)	ObjF	Time (s)	ObjF	Gap (%)
Žilina	315	158	5.17	2444	0.45	2444	0.00
	315	105	5.17	5594	0.61	5594	0.00
	315	79	5.91	8430	0.66	8430	0.00
	315	63	5.53	11125	0.66	11125	0.00
	315	32	5.33	20995	0.80	21018	0.11
	315	21	5.17	28548	1.14	28586	0.13
	315	16	5.34	34405	1.41	34425	0.06
	315	11	5.42	47092	1.91	47092	0.00
	315	8	5.63	58665	1.66	58665	0.00
	315	7	5.48	63865	2.20	63865	0.00
	315	6	5.25	69572	2.03	69572	0.00
BA	87	9	0.29	20342	0.20	20743	1.97
BB	515	52	22.88	17289	1.41	17293	0.02
KE	460	46	15.66	20042	1.47	20242	1.00
NR	350	35	7.53	22651	1.11	22651	0.00
PO	664	67	61.11	20025	2.05	20025	0.00
TN	276	28	3.55	15686	0.78	15863	1.13
TT	249	25	2.75	18873	0.84	18873	0.00

Comparison of the location-allocation approach to the radial method with $v = 20$ dividing points on the self-governing region of Žilina is reported in Table 3. The computational time is given in seconds and it contains both optimal dividing points computation and the radial weighted p-median model solving. The objective function value of the approximate solution was computed according to (14) for the resulting values of location variables y_i and the disutility matrix $\{d_{ij}\}$. The *gap* is expressed in percentage of the exact solution. Since the detailed results for other regions had similar characteristic as obtained for Žilina, we report only selected instances for the other regions. The value of parameter p in these instances was chosen so that it corresponds to the original set of problems from real emergency service system (the ratio of $|I|$ to p takes the value around 10).

The reported results prove, that the solution accuracy is very high. Even if there are some instances, which have not been solved to optimality, we can conclude that we have developed a tool for large emergency service system designing. It enables to get a good solution in a short time making use of a universal IP-solver. The possibilities of obtaining better results using different number of dividing points are discussed in [12].

The last part of computational study was aimed at exploration of the possibilities to get the optimal solution of the original problem (2)–(7) by radial formulation

(9)–(13) without the necessity of processing the whole disutility range $[d_0, d_m]$ in one turn and using only its reduced form. From the point of achieved results, this part can be considered most interesting and also most important, because the results significantly extend the applications of the radial formulation and thus considerably contribute to studied field of operations research and applied informatics.

As it was explained in Sect. 3, the main contribution of this paper consists in a two-phase algorithm. The first step searches for the maximal relevant disutility, which is used to reduce the sequence $d_0, d_1, \ldots, d_m$ for the second phase, in which the radial model (9)–(13) is solved. This innovative approach was compared to the location-allocation one from the viewpoint of computational time requirements. The achieved results are summarized in the following Table 4. Here, the value of d_{max}

Table 4 Results of numerical experiments with the two-phase radial approach for the self-governing regions of Slovakia

Region	$|I|$	p	Location-allocation			Two–phase radial approach				
			Time (s)	ObjF	d_{max}	$m+1$	Time (s)	PST (%)	ObjF	NoI
Žilina	315	158	5.17	2444	13	101	0.83	83.99	2444	6
	315	105	5.17	5594	17	115	1.27	75.52	5594	6
	315	79	5.91	8430	20	123	1.85	68.75	8430	7
	315	63	5.53	11125	20	128	1.45	73.77	11125	5
	315	21	5.17	28548	26	139	0.98	80.99	28548	1
	315	16	5.34	34405	33	141	1.34	74.85	34405	1
	315	11	5.42	47092	33	143	2.17	59.92	47092	1
	315	8	5.63	58665	42	145	2.97	47.24	58665	1
	315	7	5.48	63865	42	148	3.89	29.05	63865	1
	315	6	5.25	69572	42	149	3.83	27.09	69572	1
BA	87	9	0.29	20342	25	68	0.09	67.47	20342	1
BB	515	52	22.88	17289	26	166	1.41	93.86	17289	1
KE	460	46	15.66	20042	23	182	1.89	87.91	20042	1
NR	350	35	7.53	22651	17	118	1.25	83.39	22651	1
PO	664	67	61.11	20025	22	215	3.81	93.76	20025	1
TN	276	28	3.55	15686	30	134	4.30	−21.17	15686	7
TT	249	25	2.75	18873	24	141	0.80	71.07	18873	1
ZA	315	32	5.33	20995	26	136	0.81	84.78	20995	1
East Slo-vakia	1124	112	297.27	40713	22	222	9.11	96.94	40713	1
West Slo-vakia	1792	180	1223.69	108993	26	302	15.89	98.70	108993	1
Whole Slo-vakia	2916	273	–	–	–	477	39.52	–	161448	1

represents the maximal disutility perceived by the worst situated users. If we compare this value to the number $m + 1$ of elements included in the sequence $d_0, d_1, \ldots, d_m$, we can observe possibly high reduction of mentioned set. Since the computational time of the two-phase method is in order smaller, we computed also the percentual save of time, which is reported in columns denoted by *PST*. It must be noted, that the computational time of the radial approach covers all optimization processes of both phases. If the value of *PST* is negative, then the radial approach performed longer due to higher number of solved iterations. The number of iterations, in which the radial model (9)–(13) was solved to get the optimal solution of the original problem is given in the column *NoI*. We can observe, that suggested two-phase method is very fast thanks to simple formulations and useful features of the set-covering problem.

The reported results confirm useful features of suggested method and show, that this approach enables to solve such instances, which are not solvable when described by the location-allocation model.

5 Conclusions

This paper was aimed at special class of discrete network location problems. Since the location of centers, from which the associated service is distributed to all system users, has significant influence on transportation costs, operation efficiency and logistics performance, we focused on mastering real sized instances of the emergency service system design problem using commercial IP-solver. As it was found, common optimization techniques based on the branch and bound method often fail in solving large instances mainly when described by the location-allocation formulation. We have suggested here a composed approach to the emergency system design and explored the associated characteristics. The composition consists of combination of the "min-max" and "min-sum" approaches, where the solution techniques of both stages were based on effective radial formulation of the underlying location problems. The combination of the approaches enables to get the optimal solution of large instances by reducing the set of disutility values, which has to be processed in the second phase of suggested algorithm. Considering easy implementation of the approach on a commercial IP-solver, we have developed a flexible tool for such service system design, in which the disutility of the average user is minimized. Performed computational study has confirmed, that the two-phase approach was able to solve instance of the size of Slovak Republic whereas the location-allocation approach failed due to lack of memory. Thus we can conclude that we have constructed a very useful tool for large emergency service system designing. Presented method significantly extends the applications of the former radial approach based on the concept of dividing points, which enable to obtain good approximate solution in a short time and thus we considerably contribute to the facility location science.

Acknowledgements This paper was supported by the research grants VEGA 1/0518/15 "Resilient rescue systems with uncertain accessibility of service", VEGA 1/0463/16 "Economically efficient charging infrastructure deployment for electric vehicles in smart cities and communities", APVV-15-0179 "Reliability of emergency systems on infrastructure with uncertain functionality of critical elements".

References

1. Avella, P., Sassano, A., Vasil'ev, I.: Computational study of large scale p-median problems. Math. Program. **109**, 89–114 (2007)
2. Current, J., Daskin, M., Schilling, D.: Discrete network location models. In: Drezner, Z. et al. (ed.) Facility Location: Applications and Theory, pp. 81–118. Springer, Berlin (2002)
3. Doerner, K.F., et al.: Heuristic solution of an extended double-coverage ambulance location problem for Austria. Cent. Eur. J. Oper. Res. **13**(4), 325–340 (2005)
4. García, S., Labbé, M., Marín, A.: Solving large p-median problems with a radius formulation. INFORMS J. Comput. **23**(4), 546–556 (2011)
5. Ingolfsson, A., Budge, S., Erkut, E.: Optimal ambulance location with random delays and travel times. Heal. Care Manag. Sci. **11**(3), 262–274 (2008)
6. Janáček, J.: Approximate covering models of location problems. In: Lecture Notes in Management Science: Proceedings of the 1st International Conference on Applied Operational Research ICAOR 08, vol. 1, Sept 2008, Yerevan, Armenia, pp. 53–61 (2008)
7. Janáček, J., Kvet, M.: Approximate solving of large p-median problems. In: Operational research peripatetic post-graduate programme: Cádiz, Spain, 13–17 Sept 2011, Cádiz: Servicio de Publicaciones de la Universidad de Cádiz, 2011, pp. 221–225 (2011). ISBN: 978-84-9828-348-8
8. Janáček, J., Kvet, M.: Public service system design with disutility relevance estimation. In: Proceedings of the 31st International Conference Mathematical Methods in Economics, 11–13 Sept 2013, Jihlava, Czech Republic, pp. 332–337 (2013). ISBN: 978-80-87035-76-4
9. Janáček, J., Kvet, M.: Relevant network distances for approximate approach to large p-median problems. In: Operations Research Proceedings 2012: Selected Papers of the International Conference on Operations Research: 4–7 Sept 2012, pp. 123–128, Springer, Hannover, Germany (2014). ISSN 0721-5924, ISBN 978-3-319-00794-6,
10. Jánošíková, Ľ.: Emergency medical service planning. Commun.—Sci. Lett. Univ Žilina **9**(2), 64–68 (2007)
11. Kvet, M., Janáček, J.: Min-max optimal public service system design. Croat. Oper. Res. Rev. **6**(1), 17–27 (2015)
12. Kvet, M., Kvet, M.: Accuracy sensitivity of the radial approach to large public service system design. In: CISTI 2016: Actas de la 11a Conferencia Ibérica de Sistemas y Tecnologías de Información, 2016, pp. 128–135 (2016). ISBN: 978-989-98434-6-2
13. Kvet, M., Matiaško, K.: Magnetic resonance imaging results processing: brain tumour marker value processing. In: Digital Technologies 2013: 29–31 May 2013, pp. 149-159, University of Žilina, Žilina (2013). ISBN: 978-80-554-0682-4
14. Kvet, M., Matiaško, K.: Epsilon temporal data in MRI results processing. Digital Technologies 2014: 9–11 July 2014, pp. 209–217. University of Žilina, Žilina (2014). ISBN 978-1-4799-3301-3
15. Kvet, M., Matiaško, K., Vajsová, M.: Sensor based transaction temporal database architecture. In: World Conference on Factory Communication Systems (WFCS): Communication in Automation, 27–29 May 2015, Palma de Mallorca, pp. 1-8, (2015). ISBN: 978-1-4799-8243-1
16. Marianov, V., Serra, D.: Location problems in the public sector. In: Drezner, Z. et al. (ed.) Facility Location: Applications and Theory, pp. 119-150. Springer, Berlin (2002)

Verbal Decision Analysis Applied to the Prioritization of Influencing Factors in Distributed Software Development

Marum Simão Filho, Plácido Rogério Pinheiro and Adriano Bessa Albuquerque

Abstract A project manager of distributed software development faces many challenges. Some of them relate to task allocation among remote teams. To assign a task to team, the project manager takes into account several factors such as project manager maturity and team availability. Most of the time, the manager takes the decision subjectively. The verbal decision analysis is an approach to solve problems through multi-criteria qualitative analysis, i.e., it considers the analysis of subjective criteria. This paper describes the application of verbal decision analysis method ZAPROS III-i to prioritize the factors that the project managers should take into account when allocating tasks in projects of distributed software development.

Keywords Distributed software development · Task allocation · Verbal decision analysis · ZAPROS III-i

1 Introduction

Software development companies are increasingly investing in distributed software development. As main reasons for this, we mention the expansion of the workforce capacity, the prospect of gaining new customers around the world and cost reduction expectation [25]. However, working in a distributed manner introduces several challenges, such as language and time zone differences and increased

M. Simão Filho (✉) · P.R. Pinheiro · A.B. Albuquerque
University of Fortaleza, Fortaleza, Brazil
e-mail: marumsimao@gmail.com; marum@fa7.edu.br

P.R. Pinheiro
e-mail: placido@unifor.br

A.B. Albuquerque
e-mail: adrianoba@unifor.br

M. Simão Filho
7 de Setembro College, Fortaleza, Brazil

© Springer International Publishing AG 2018
Á. Rocha and L.P. Reis (eds.), *Developments and Advances in Intelligent Systems and Applications*, Studies in Computational Intelligence 718,
DOI 10.1007/978-3-319-58965-7_4

49

complexity of coordinating and controlling projects [23]. The allocation of task is, by nature, a complex activity. Besides, it is critical to the success of the project [22]. If we consider working with distributed teams, task allocation becomes even more critical due to the lack of knowledge on the factors that influence decisions on task allocation [8].

The assignment of tasks to the teams is regarded as a decision problem. Commonly, the project manager makes this decision based on his/her experience and knowledge about the project and the teams involved. In this case, a high degree of subjectivity is present in the decision-making process. This sort of situation is appropriate for Verbal Decision Analysis (VDA). VDA is an approach based on multi-criteria problem solving through its qualitative analysis [16], i.e., VDA methods take into consideration the criteria's subjectivity.

This paper describes the application of a methodology using a VDA method to set a prioritization scale on the factors to be considered by project managers when allocating tasks in projects of distributed software development. First, we conducted interviews with experts to identify the criteria and the criteria values that were used by the model. Then, we applied a questionnaire to a group of project managers aiming to characterize each factor through the criteria and their criteria values. Finally, the ZAPROS III-i method was applied to rank order the factors.

The remainder of the paper is organized as follows. Section 2 presents some issues involving task allocation in distributed software development. Section 3 provides an overview of the VDA approach and describes the ZAPROS III-i method. Section 4 details the application of the ZAPROS III-i method to the problem of prioritizing influencing factors in task allocation in distributed software development. Section 5 presents the results of this research. In the end, Sect. 6 provides the conclusions and suggestions for further work.

2   Task Allocation in Distributed Software Development

The allocation of tasks is a critical activity for any kind of project, especially in a distributed scenario. Most of the time, few factors drive the allocation of tasks, such as hand labor costs. Risks and other relevant factors such as the workforce skills, innovation potential of different regions, or cultural factors are often insufficiently recognized [11].

Many studies about the tasks allocation in DSD have been carried out along the years aiming at mapping this topic and its features. Lamersdorf et al. [11] developed an analysis of the existing approaches to distribution of duties. The analysis was comprehensive and involved procedures for the distributed development, distributed generation, and distributed systems areas. Lamersdorf et al. [12] conducted a survey on the state of practice in DSD in which they investigated the criteria that influence task allocation decisions. Lamersdorf and Münch [10] presented TAMRI (Task Allocation based on Multiple cRIteria), a model based on multiple criteria and influencing factors to support the systematic decision of task allocation in distributed development projects. Ruano-Mayoral et al. [29] presented a methodological framework to allocate work packages among participants in global software development projects.

Marques et al. [23] performed a systematic mapping, which enabled us to identify models that propose to solve the problems of allocation of tasks in DSD projects. Marques et al. [22] also performed a tertiary review applying the systematic review method on systematic reviews that address the DSD issues. Galviņa and Šmite [9] provided an extensive literature review for understanding the industrial practice of software development processes and concluded that the evidence of how these projects are organized is scarce. Babar and Zahedi [3] presented a literature review considering the studies published in the International Conference in Global Software Engineering (ICGSE) between 2007 and 2011. They found that the vast majority of the evaluated studies were in software development governance and its sub-categories, and much of the work had focused on the human aspects of the GSD rather than technical aspects.

Almeida et al. [1] presented a multi-criteria decision model for planning and fine-tuning such project plans: Multi-criteria Decision Analysis (MCDA). The model was developed using cognitive mapping and MACBETH (Measuring Attractiveness by a Categorical Based Evaluation Technique) [4]. In [2], Almeida et al. applied (MCDA) on the choice of DSD Scrum project plans that have a better chance of success.

Simão et al. [30] conducted a quasi-systematic review of studies of task allocation in DSD projects that incorporate agile practices. The study brought together a number of other works, allowing the establishment of the many factors that influence the allocation of tasks in DSD, which we can highlight: technical expertise, expertise in business, project manager maturity, proximity to client, low turnover rate of remote teams, availability, site maturity, personal trust, time zone, cultural similarities, and willingness at site. These factors are very important for this work and will be used later.

3   Verbal Decision Analysis

Decision-making is an activity that is part of people's and organizations' lives. In most problems, to make a decision, a situation is assessed against a set of characteristics or attributes, i.e., it involves the analysis of several factors, also called criteria. When a decision can generate a considerable impact, such as management decisions, and must take into account some factors, the use of methodologies to support the decision-making process is suggested, because choosing the inappropriate alternative can lead to waste of resources, time, and money, affecting the company.

The decision-making scenario that involves the analysis of alternatives from several viewpoints is called multi-criteria decision analysis and is supported by multi-criteria methodologies [4]. These methodologies favor the generation of knowledge about the decision context, which helps raise the confidence of the decision maker [8, 21].

The verbal decision analysis is an approach to solving multi-criteria problems through qualitative analysis [14]. The VDA supports the decision-making process through the verbal representation of problems. Some examples of the application of VDA in real problems are given next. In [19], Machado et al. applied VDA in selecting specific practices of CMMI. In [20], Machado applied VDA for selecting approaches of project management. In [24, 41], Mendes et al. and Tamanini et al., respectively, used VDA in digital TV applications. In [38], Tamanini et al. proposed a VDA-based model to cashew chestnut industrialization process. In [39, 40], Tamanini et al. developed studies applying VDA to the diagnosis of Alzheimer's disease. VDA applied to Alzheimer's disease also was approached in [5–7]. In [42], Tamanini and Pinheiro approached the incomparability problem on ZAPROS method. In [28], Pinheiro et al. applied VDA on the choice of educational tools prototypes.

According to [26], the VDA framework is based on the same principles as Multi-Attribute Utility Theory (MAUT). However, it applies a verbal process for preference elicitation and evaluation of alternatives without converting these verbal values to numbers. The traditional methods of VDA aimed at solving problems with many alternatives and a limited number of criteria and criteria values, since they were designed for the construction of a general rule for the decision, regardless of which alternatives belonged to the real alternatives set. However, this characteristic has changed recently, and new methods that elicit the preferences based on the real alternatives to the problem have been proposed.

The VDA methodologies can be used for ordering or sorting the alternatives. Among the classification methods, we can mention ORCLASS, SAC, DIFCLASS, and CYCLE. Some sorting methods are PACOM, ARACE, and those from ZAPROS family (ZAPROS-LM, STEPZAPROS, ZAPROS III and III-i) [35]. Figure 1 shows the VDA ordering methods.

3.1   The ZAPROS III-I Method for Rank Ordering

The ZAPROS methodology aims at ranking multi-criteria alternatives in scenarios involving a rather small set of criteria and criteria values, and a great number of alternatives. It is structured in three stages: Problem Formulation, Elicitation of Preferences and Comparison of Alternatives.

In ZAPROS LM method, we carry out the elicitation of preferences by comparing vectors of alternatives [16]. These vectors are composed of criteria values. They differ from only two values at the same time. Then, using a pair of vectors, we can structure the scale of preferences according to the decision maker's answers. After that, we will define the real alternatives of the problem and the values of criteria that they represent. In the end, we obtain the value of each alternative based

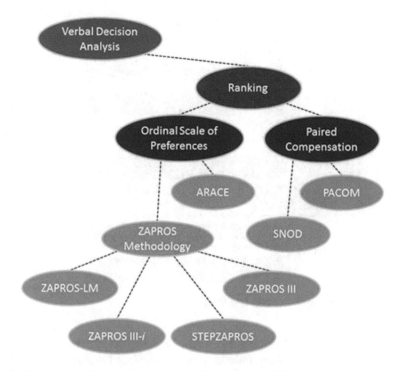

Fig. 1 VDA methods for ordering alternatives (Adapted from [35])

on the criteria values weights defined in the scale of preferences and, then, we can rank order the alternatives.

The ZAPROS III method [13] is an evolution of the ZAPROS LM. There are modifications that make it more efficient and more accurate on inconsistencies. ZAPROS III introduces the concept of Quality Variations (QV), which represents the distances between the evaluations of two criteria, and uses these values to make the preferences elicitation. Furthermore, it uses the Formal Index of Quality (FIQ) to rate the alternatives set and, hence, minimize the number of pairs of alternatives that need to be compared to obtain the problem's result. FIQ estimates the quality degree of an alternative through the comparison of it to the First Reference Situation of the problem. The highest the difference between these two alternatives (and, consequently, the degree presented on the alternative's FIQ), the less preferable this would be. This way, one can notice that the FIQ of the first reference situation will always be equal to zero.

The subjectivity and the qualitative aspect of ZAPROS method can cause losses to the method's comparison capacity and make the incomparability cases between the alternatives unavoidable [15]. ZAPROS III-i introduces modifications in the comparison process of alternatives so that it could minimize or even eliminate the incomparability problem of the ZAPROS method. More details about the modification on ZAPROS III-i can be found in [43].

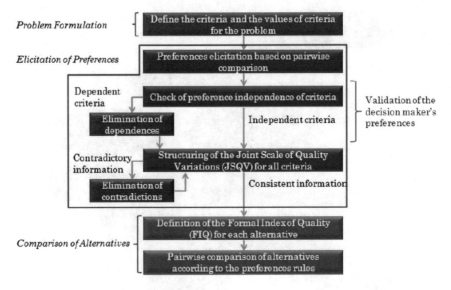

Fig. 2 Procedure to apply ZAPROS III-i Method [36]

Figure 2 shows the ZAPROS III-i method's flowchart to rank order a set of alternatives. The methodology follows the same problem formulation proposed in [13]. In the first stage, we obtain the relevant criteria and their values to the decision-making process. The criteria values must be sorted in an ascending order of preference (from most to least preferable). In case of three criteria and three criteria values, we have the following criteria values for each criterion: A1, A2 and A3 (for criterion A), B1, B2 and B3 (for criterion B), and C1, C2 and C3 (for criterion C). A1, B1 and C1 are the most preferable values, and A3, B3 and C3 are the least preferable values for criteria A, B and C, respectively. In this case, possible alternatives are represented as (A1, B2, C3) and (A3, B1, C2). In the second stage, we generate the scale of preferences based on the decision maker's preference. In the last stage, we perform the comparison between the alternatives based on the decision maker's preferences. More details about this process can be found in [13, 27, 36].

4 The Application of the Methodology

To rank order the factors that project managers should consider when allocating tasks in distributed software development projects, we applied a methodology consisting of four main steps: A—Identification of the Alternatives (the factors); B —Definition of the Criteria and the Criteria Values; C—Characterization of the Alternatives; and D—The ZAPROS-III-i Method Application [32]. A flowchart of the methodology can be seen in Fig. 3. The steps are explained next.

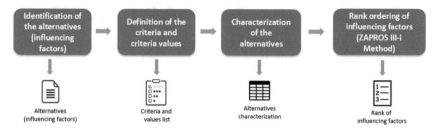

Fig. 3 Flowchart of the methodology

A *Identification of the Alternatives*

First, we conducted a literature research to identify the main influencing factors that should be considered when allocating tasks in distributed software development projects. They serve as alternatives to our decision problem [31]. Table 1 shows the factors found as result of this research.

B *Definition of the Criteria and the Criteria Values*

Next, we interviewed a group of 4 project management experts in order to define the criteria and the criteria values. This is the definition stage of the criteria. For each criterion, we established a scale of values associated with it [17, 18, 41].

The criteria values were ordered from the most preferable value to the least preferable one. As result of this step, we got the list of criteria and criteria values for the problem of selecting the most important factors to be considered in task allocation in DSD projects, which is listed next [31]:

1. Criterion A: Facility for carrying out the task remotely, i.e., how much easier it becomes to implement the remote task if the factor is present.

   • A1. It facilitates much: The implementation of the remote task is much easier if the factor is present.

Table 1 Influencing factors on task allocation in DSD projects

ID	Alternatives
Factor1	Technical expertise
Factor2	Expertise in business
Factor3	Project manager maturity
Factor4	Proximity to client
Factor5	Low turnover rate
Factor6	Availability
Factor7	Site maturity
Factor8	Personal trust
Factor9	Time zone
Factor10	Cultural similarities
Factor11	Willingness at site

- A2. It facilitates: The implementation of the remote task is easier if the factor is present.
- A3. Indifferent: The presence of the factor is indifferent to the implementation of the remote task.

2. Criterion B: Time for the project.

- B1. High gain: The presence of the factor can cause much reduction of the period referred to perform the task.
- B2. Moderate gain: The presence of the factor may cause some reduction of the time limit for performing the task.
- B3. No gain: The presence of the factor does not cause changes to the deadline to execute the task.

3. Criterion C: Cost for the project.

- C1. High gain: The presence of the factor can cause a lot of cost reduction expected to perform the task.
- C2. Moderate gain: The presence of the factor may cause some reduction of the time limit for performing the task.
- C3. No gain: The presence of factor induces no change compared to the estimated cost to perform the task.

C *Characterization of the Alternatives*

We created a questionnaire to gather information and opinions about the factors that influence the allocation of tasks in DSD projects. We applied the questionnaire to the Web to a group of 20 project managers and consisted of two parts. The first part aimed to trace the respondents profile about his/her professional experience and education.

The respondents' profiles can be summarized as follows. 30% of respondents have bachelor's degree, 60% have master's degrees and 10% are doctors or higher. All respondents work with software development for over 8 years. 40% work in private companies and 60% in public companies. 65% work in companies whose business is to provide IT services whereas 35% do not. 40% of respondents have between 4 and 8 years of experience in managing software development projects whereas 60% have more than 8 years. 65% of participants are certified in project management field, while 35% do not have any certification. 80% of respondents managed more than 8 software development projects and 20% managed less than 8 projects. Considering projects of distributed development of software, 20% of respondents managed more than 8 projects, 55% managed from 1 to 8 projects, and 25% did not manage any project.

The second part of the questionnaire inquired the views of experts on the factors that influence the allocation of tasks in DSD projects. As explained, for our problem, we described such influencing factors as alternatives (see Table 1). Thus, in every question, the professional analyzed the influencing factors about a set of

Table 2 List of criteria and criteria values with description

Criteria	Criteria values	Description
A: Facility for carrying out the task remotely	Al. It facilitates much	The implementation of the remote task is much easier if the factor is present
	A2. It facilitates	The implementation of the remote task is easier if the factor is present
	A3. Indifferent	The presence of the factor is indifferent to the implementation of the remote tasik
R: Time for the project	Bl. High gain	The presence of the factor can cause much reduction of the period referred to perform the task
	B2 Moderate gain	The presence of the factor may cause some reduction of the time limit for performing the task
	B3. No gain	The presence of the factor does not cause changes to the deadline to execute the tasik
C: Cost for the project	CI. High gain	The presence of the factor can cause a lot of cost reduction expected to perform the task
	C2. Moderate gain	The presence of the factor may cause some reduction of the time limit for performing the task
	C3. No gain	The presence of factor induces no change compared to the estimated cost to perform the task

criteria and criteria values (shown in Table 2) and selected what criterion value that best fitted the factor analyzed.

An example of question is as follows:

1. Factor: Expertise in business—knowledge of the team about the client's business:

   (a) Criterion A: Facility for carrying out the task remotely
       ($\cdot$) A1. It facilitates much. ($\cdot$) A2. It facilitates. ($\cdot$) A3. Indifferent.
   (b) Criterion B: Time for the project
       ($\cdot$) B1. High gain. ($\cdot\cdot\cdot$) B2. Moderate gain. ($\cdot\cdot\cdot$) B3. No gain.
   (c) Criterion C: Cost for the project
       ($\cdot\cdot\cdot$) B1. High gain. ($\cdot\cdot\cdot$) B2. Moderate gain. ($\cdot\cdot\cdot$) B3. No gain.

We did the same for the other ten factors. Then, we analyzed the responses to determine the criteria values representing the alternatives. For each influencing factor, we filled the final table based on the replies of the majority of professionals. We then selected the value of the criterion that had the greatest number of choices

Table 3 Characterization of alternatives according to answers collected in the questionnaire [31]

Criteria/alternatives	Facility for carrying out the task remotely			Time for the project			Cost for the project			
	Al	A2	A3	Bl	B2	B3	C1	C2	C3	Final vector
Factor1	**11**	7	2	**13**	6	1	**11**	7	2	A1B1C1
Factor2	**15**	3	2	**13**	7	0	**10**	8	2	A1B1C1
Factor3	8	**11**	1	5	**14**	1	7	**10**	3	A2B2C2
Factor4	**13**	4	3	8	**10**	2	8	**10**	2	A1B2C2
Factor5	**14**	6	0	**15**	4	1	**12**	7	1	A1B1C1
Factor6	**10**	8	2	**13**	5	2	**9**	6	5	A1B1C1
Factor7	**16**	3	1	**11**	9	0	9	**11**	0	A1B1C2
Factor8	8	**10**	2	6	**11**	3	3	**13**	4	A2B2C2
Factor9	3	**12**	5	3	8	**9**	3	6	**11**	A2B3C3
Factor 10	4	**13**	3	3	**10**	7	3	8	**9**	A2B2C3
Factor 11	**10**	8	2	**10**	8	2	**9**	6	5	A1B1C1

to represent the alternative. Table 3 summarizes the responses to the questionnaire, showing the sum of the answers and characterization of alternatives according to the values of each criterion (the "Final Vector" column). The bold numbers in gray cells in the table indicate the criteria values selected by most of the interviewed professionals to represent a certain factor.

We emphasize that the various answers given by professionals, considering they have experienced project managers, were related to the fact that they have different professional backgrounds. Thereby, the characterization of a particular factor was based on answers given by most professionals.

D *The ZAPROS-III-i Method Application*

After defining and characterizing the alternatives, we moved on to the stage of ordering. At this stage, we applied the ZAPROS III-i method to put in order the influencing factors, such that it is possible to establish a ranking of them.

In order to facilitate the decision-making process and perform it consistently, we used the ARANAÚ tool, presented in [37, 41, 43]. The tool, which was implemented in Java platform, was first developed in [37] to support ZAPROS III method. In this work, we used the updated version to ZAPROS III-i method. The use of ZAPROS III-i method in the ARANAÚ tool requires four steps, as follows: 1. Criteria and criteria values definition; 2. Preferences elicitation; 3. Alternatives definition; and 4. Results generation.

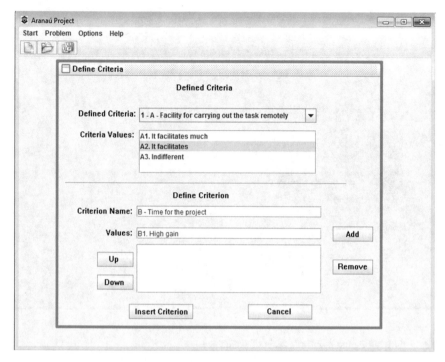

Fig. 4 Introducing criteria into the ARANAÚ tool

The process runs as follows. First, we introduced the criteria presented in the problem into the ARANAÚ tool, as shown in Fig. 4. Next, the decision-maker decides the preferences. The interface for elicitation of preferences presents questionings that can be easily answered by the decision-maker to obtain the scale of preferences. The process occurs in two stages: elicitation of preferences for quality variation of the same criteria, and elicitation of preferences between pairs of criteria. The questions provided require a comparison considering the two reference situations [36]. Figure 5 exposes the ARANAÚ's interface for comparison of quality variations.

Once the scale of preferences is structured, the next step is to define the problem's alternatives. The alternatives to our problem are the factors shown in Subsection IV-0. This step is performed through the interface of ARANAÚ shown in Fig. 6.

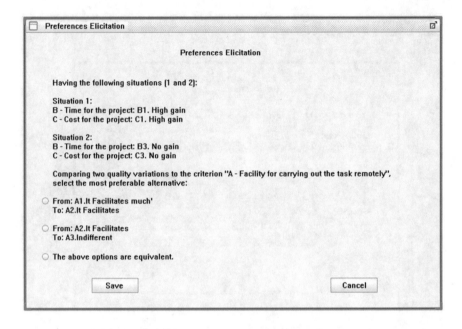

Fig. 5 Structuring the problem's scale of preferences

5 Results

After introducing all the data and answering the questions, the decision maker is presented with the result in a table containing the alternatives and their criteria evaluations, formal index of quality and rank, as exposed in Table 4. Note that there are five alternatives (factors) that are in the same ranking position (first position), and their FIQ's values are equals to zero. This occurs because all of them got the best evaluation according to the survey filled out by the professionals (A1, B1, C1), which is the best possible evaluation, as explained in Subsection 01.

A graph showing the dominance relations between the alternatives is also generated by the ARANAÚ tool and is exposed to provide a more detailed analysis of the problem's resolution. This graph can be seen in Fig. 7.

Further research involving the application of a hybrid methodology using VDA methods to initially classify (ORCLASS) and then rank order (ZAPROS III-i) the most relevant factors to be considered by project managers when allocating tasks in projects of distributed development of software can be found in [33]. A proposal of a model for task allocation in projects of distributed software development based on VDA methods ORCLASS and ZAPROS III-I can be found in [34].

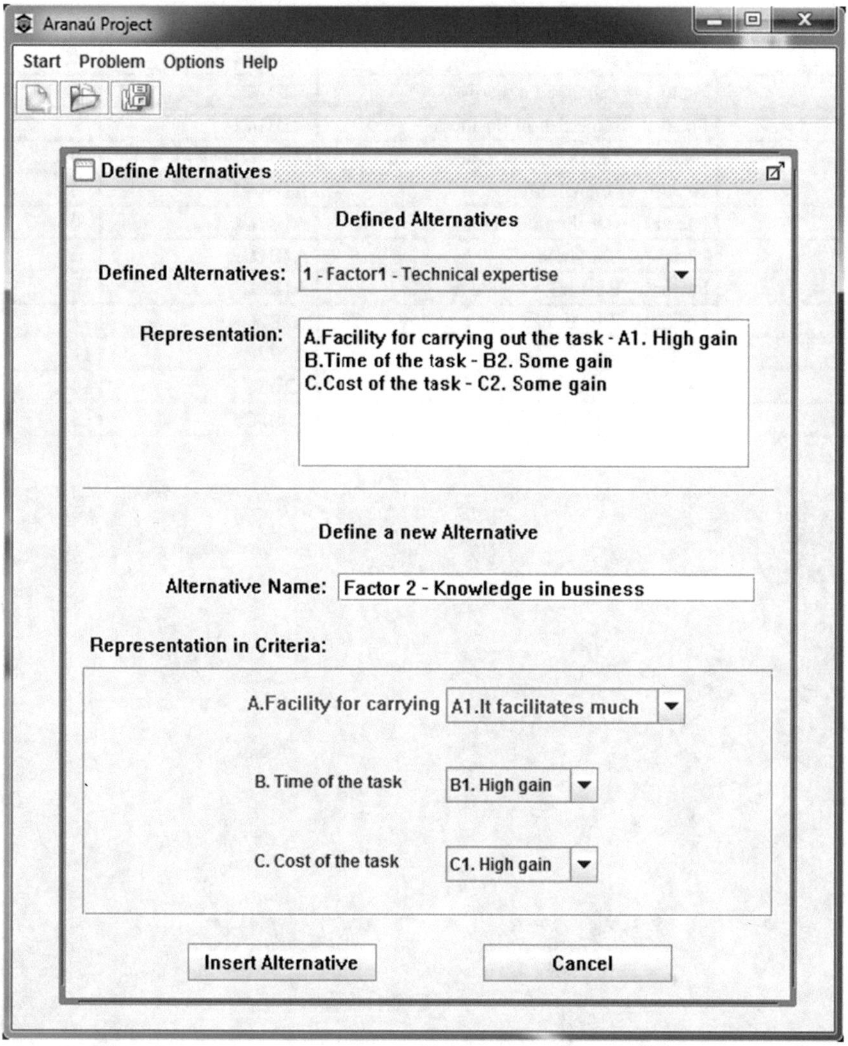

Fig. 6 Defining the problem's alternatives

Table 4 The final ranking of alternatives

Rank	Alternative	Representation	FIQ
1	Factor1—Technical expertise	A1B1C1	0
1	Factor2—Expertise in business	A1B1C1	0
1	Factor5—Low turnover rate	A1B1C1	0
1	Factor6—Availability	A1B1C1	0
1	Factor11—Willingness at site	A1B1C1	0
2	Factor7—Site maturity	A1B1C2	6
3	Factor4—Proximity to client	A1B2C2	10
4	Factor3—Project manager maturity	A2B2C2	11
4	Factor8—Personal trust	A2B2C2	11
5	Factor10—Cultural Similarities	A2B2C3	14
6	Factor9—Time zone	A2B3C3	18

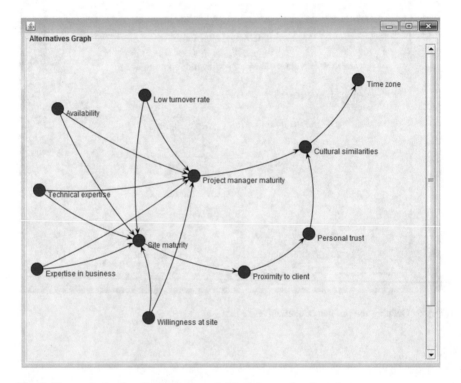

Fig. 7 The graph showing the dominance relations between the alternatives

6 Conclusion and Future Works

For large software companies, working with distributed teams has been an alternative increasingly present in projects of software development. However, the allocation of tasks among remote teams is very critical since there are many factors that project managers should take into consideration. Typically, this multi-criteria decision-making problem involves subjective aspects. The verbal decision analysis methods support decision-making process through multi-criteria qualitative analysis.

The main contribution of this work was to apply a methodology based on VDA method ZAPROS III-i to prioritize the factors that project managers should consider when allocating tasks to distributed teams. A tool called ARANAÚ supported this work allowing performing the tasks in a fast and practical way. Initially, we conducted interviews and applied questionnaires to a group of project management experts so that we could identify the alternatives (the influencing factors), define the criteria and their respective values, and characterize the alternatives to use in the method. Another contribution of this work was to provide information about factors that influence the allocation of tasks to project managers of distributed software development.

As future work, we intend to combine the ZAPROS III-i method to other decision support methods so that we can apply them in the allocation of specific software engineering tasks. Another future work is the application of the factors arising from this research in actual projects in order to validate the model's result. Finally, we intend to apply VDA methods to help choosing the team that should be assigned a specific task, based on the task characteristics and teams profiles.

Acknowledgements The first author is thankful for the support given by the Coordination for the Improvement of Higher Level-or Education- Personnel (CAPES) and 7 de Setembro College during this project. The second author is grateful to National Counsel of Technological and Scientific Development (CNPq) via Grants #305844/2011-3. The authors would like to thank University of Fortaleza for all the support.

References

1. Almeida, L.H., Albuquerque, A.B., Pinheiro, P.R.: A multi-criteria model for planning and fine-tuning distributed scrum projects. In: 6th IEEE International Conference on Global Software Engineering (2011)
2. Almeida, L.H., Albuquerque, A.B., Pinheiro, P.R.: Applying multi-criteria decision analysis to global software development with scrum project planning. Lect. Notes Comput. Sci. **6954**, 311–320 (2011)
3. Babar, M.A., Zahedi, M.: Global software development: a review of the state-of-the-art (2007–2011). IT University Technical report Series. IT University of Copenhagen (2012)
4. Bana e Costa, C.A., Sanchez-Lopez, R., Vansnick, J.C., De Corte, J.M.: Introducción a MACBETH. In: Leyva López, J.C. (ed.) Análisis Multicriterio para la Toma de Decisiones: Métodos y Aplicaciones, Plaza y Valdés, México, pp. 233–241 (2011)

5. Castro, A.K.A., Pinheiro, P.R.: Pinheiro MCD (2007) Applying a decision making model in the early diagnosis of alzheimer's disease. Rough Sets Knowl. Technol. Lect. Notes Comput. Sci. **4481**, 149–156 (2007)
6. Castro, A.K.A., Pinheiro, P.R., Pinheiro, M.C.D.: A multicriteria model applied in the diagnosis of alzheimer's disease. Rough Sets Knowl. Technol. Lect. Notes Comput. Sci. **5009**, pp. 612–619 (2008)
7. Castro, A.K.A., Pinheiro, P.R., Pinheiro, M.C.D., Tamanini, I.: Applied hybrid model in the neuropsychological diagnosis of the alzheimer's disease: a decision making study case. Int. J. Soc. Humanist. Comput. (IJSHC) **1**, 3:331–345 (2010). doi:10.1504/IJSHC.2010.032692
8. Evangelou, C., Karacapilidis, N., Khaled, O.A.: Interweaving knowledge management, argumentation and decision making in a collaborative setting: the KAD ontology model. Int. J. Knowl. Learn. **1**, 1/2, 130–145 (2005)
9. Galviņa, Z., Šmite, D.: Software development processes in globally distributed environment. In: Scientific Papers, University Of Latvia, vol. 770. Computer Science and Information Technologies (2011)
10. Lamersdorf, A., Münch, J.: A multi-criteria distribution model for global software development projects. Brazilian Comput. Soc. (2010)
11. Lamersdorf, A., Münch. J., Rombach, D.: Towards a multi-criteria development distribution model: an analysis of existing task distribution approaches. In: IEEE International Conference on Global Software Engineering, ICGSE 2008
12. Lamersdorf, A., Münch, J., Rombach, D.: A survey on the state of the practice in distributed software development: criteria for task allocation. In: 4th IEEE International Conference on Global Software Engineering, ICGSE (2009)
13. Larichev, O.I.: Ranking multicriteria alternatives: the method ZAPROS III. Eur. J. Oper. Res. **131**, 3:550–558 (2001)
14. Larichev, O.I., Brown, R.: Numerical and verbal decision analysis: comparison on practical cases. J. Multicrit. Decis. Anal. **9**(6), 263–273 (2000)
15. Larichev, O.: Method ZAPROS for multicriteria alternatives ranking and the problem of incomparability. Informatica **12**(1), 89–100 (2001)
16. Larichev, O.I., Moshkovich, H.M.: Verbal Decision Analysis for Unstructured Problems. Kluwer Academic Publishers, Boston (1997)
17. Machado, T.C.S., Menezes, A.C., Pinheiro, L.F.R., Tamanini, I., Pinheiro, P.R.: The selection of prototypes for educational tools: an applicability in verbal decision analysis. In: IEEE International Joint Conferences on Computer, Information, and Systems Sciences, and Engineering (2010)
18. Machado, T.C.S., Menezes, A.C., Pinheiro, L.F.R., Tamanini, I., Pinheiro, P.R.: Applying verbal decision analysis in selecting prototypes for educational tools. In: IEEE International Conference on Intelligent Computing and Intelligent Systems, Xiamen, China, pp. 531–535 (2010)
19. Machado, T.C.S., Pinheiro, P.R., Albuquerque, A.B., de Lima, M.M.L.: Applying verbal decision analysis in selecting specific practices of CMMI. Lect. Notes Comput. Sci. **7414**, 215–221 (2012)
20. Machado, T.C.S.: Towards aided by multicriteria support methods and software development: a hybrid model of verbal decision analysis for selecting approaches of project management. Master Thesis. Master Program in Applied Computer Sciences, University of Fortaleza (2012)
21. Machado, T.C.S., Pinheiro, P.R., Tamanini, I.: Project management aided by verbal decision analysis approaches: a case study for the selection of the best SCRUM practices. Int. Trans. Oper. Res. **22**, 2:287–312 (2014). doi:10.1111/itor.12078
22. Marques, A.B., Rodrigues, R., Conte, T.: Systematic literature reviews in distributed software development: a tertiary study. IEEE Int. Conf. Global Softw. Eng. ICGSE **2012**, 134–143 (2012)
23. Marques, A.B., Rodrigues, R., Prikladnicki, R., Conte, T.: Alocação de Tarefas em Projetos de Desenvolvimento Distribuído de Software: Análise das Soluções Existentes.

In: II Brazilian Conference on Software, V WDDS—Workshop on Distributed Software Development, São Paulo (2011)

24. Mendes, M.S., Carvalho, A.L., Furtado, E., Pinheiro, P.R.: A co-evolutionary interaction design of digital TV applications based on verbal decision analysis of user experiences. Int. J. Digit. Cult. Electron. Tourism 1, 312–324 (2009)

25. Miller, A.: Distributed agile development at microsoft patterns & practices. Microsoft Patterns Pract. (2008)

26. Moshkovich, H.M., Mechitov, A.: Verbal decision analysis: foundations and trends. Adv. Decis. Sci. 2013, Article ID 697072, 9 pages (2013). doi:10.1155/2013/697072

27. Moshkovich, H.M., Mechitov, A., Olson, D.: Ordinal judgments in multiattribute decision analysis. Eur. J. Oper. Res. 137(3), 625–641 (2002)

28. Pinheiro, P.R., Machado, T.C.S., Tamanini, I.: Verbal decision analysis applied on the choice of educational tools prototypes: a study case aiming at making computer engineering education broadly accessible. Int. J. Eng. Educ. 30, 585–595 (2014)

29. Ruano-Mayoral, M., Casado-Lumbreras, C., Garbarino-Alberti, H., Misra, S.: Methodological framework for the allocation of work packages in global software development. J. Softw. Evol. Proc. 26, 476–487 (2013). doi:10.1002/smr.1618

30. Simão, F.M., Pinheiro, P.R., Albuquerque, A.B.: Task allocation approaches in distributed agile software development: a quasi-systematic review. In: Proceedings of the 4th Computer Science On-line Conference 2015 (CSOC2015), Software Engineering in Intelligent Systems Series, Zlín, vol. 3, pp. 243–252 (2015). doi:10.1007/978-3-319-18473-9_24

31. Simão, F.M., Pinheiro, P.R., Albuquerque, A.B.: Task allocation in distributed software development aided by verbal decision analysis. In: Proceedings of the 5th Computer Science On-line Conference 2016 (CSOC2016), Software Engineering Perspectives and Application in Intelligent Systems Series, Zlín, vol. 2, pp. 127–137 (2016). doi:10.1007/978-3-319-33622-0_12

32. Simão, F.M., Pinheiro, P.R., Albuquerque, A.B.: Applying verbal decision analysis in distributed software development-rank ordering the influencing factors in task allocation. In: Proceedings of the 11th Iberian Conference on Information Systems and Technologies (CISTI'2016), Gran Canaria, España, vol. I, pp. 205–210 (2016)

33. Simão, F.M., Pinheiro, P.R., Albuquerque, A.B.: Applying verbal decision analysis to task allocation in distributed software development. In: Proceedings of the 28th International Conference on Software Engineering & Knowledge Engineering, San Francisco, pp. 402–407 (2016). doi:10.18293/SEKE2016-181

34. Simão, F.M., Pinheiro, P.R., Albuquerque, A.B.: Towards a model for task allocation in distributed software development. In: Proceedings of the VII Brazilian Conference on Software (CBSoft 2016), X Workshop on Distributed Software Development, Ecosystem Software and Systems-of-Systems (WDES), Maringá (2016)

35. Tamanini, I.: Hybrid approaches of verbal decision analysis methods. Doctor Thesis, Graduate Program in Applied Computer Science, University of Fortaleza (2014)

36. Tamanini, I.: Improving the ZAPROS method considering the incomparability cases. Master Thesis, Graduate Program in Applied Computer Sciences, University of Fortaleza (2010)

37. Tamanini, I.: Uma ferramenta Estruturada na Análise Verbal de Decisão Aplicando ZAPROS. Computer Sciences. University of Fortaleza (2007)

38. Tamanini, I., Carvalho, A.L., Castro, A.K.A., Pinheiro, P.R.: A novel multicriteria model applied to cashew chestnut industrialization process. Adv. Soft Comput. 58(1), 243–252 (2009)

39. Tamanini, I., de Castro, A.K.A., Pinheiro, P.R.: Pinheiro MCD (2009) Towards an applied multicriteria model to the diagnosis of Alzheimer's disease: a neuroimaging study case. IEEE Int. Conf. Intell. Comput. Intell. Syst. 3, 652–656 (2009)

40. Tamanini, I., de Castro, A.K.A., Pinheiro, P.R., Pinheiro, M.C.D.: Verbal decision analysis applied on the optimization of Alzheimer's disease diagnosis: a study case based on neuroimaging. Adv. Exp. Med. Biol. 696, 555–564 (2011)

41. Tamanini, I., Machado, T.C.S., Mendes, M.S., Carvalho, A.L., Furtado, M.E.S., Pinheiro, P.R.:
    A model for mobile television applications based on verbal decision analysis. Adv. Comput.
    Innov. Inf. Sci. Eng. **1**(1), 399–404 (2008)
42. Tamanini, I., Pinheiro, P.R.: Challenging the incomparability problem: an approach
    methodology based on ZAPROS. Model. Comput. Optim. Inf. Syst. Manag. Sci. Commun.
    Comput. Inf. Sci. **14**, 338–347 (2008)
43. Tamanini, I., Pinheiro, P.R.: Reducing Incomparability in Multiciteria Decision Analysis: An
    Extension of the ZAPROS Methods. Pesquisa Operacional, vol. 31, n. 2, pp. 251–270 (2011).
    doi:10.1590/S0101-74382011000200004

Agile Documentation Tool Concept

Stefan Voigt⊙, Detlef Hüttemann, Andreas Gohr and Michael Große

Abstract Documentation is often neglected in agile software projects, even if software developers perceive a need for good documentation. One reason can be found in improper documentation tools. This paper provides an overview of the central conceptual ideas for an agile documentation tool.

1 Introduction

Agile software development methods play an important role in research and practice [2, 6, 8]. "Traditionalists" and "agilists" have debated the correct interpretation of values and principles in the Agile Manifesto [11]. The second value "Working software over comprehensive documentation" is frequently misunderstood. Some agilists even see documentation as a waste of time since it does not contribute to the final product [21]. Agile methods and documentation are not contradictory, though [14, 22]. A certain amount of documentation is essential [14, 24]. Even agile developers consider documentation to be an important issue [26]. We see a need for techniques and methods that support documentation in agile environments [5]. Documentation has to become more easily writable, manageable and updatable [18]. This paper will outline a concept for an agile documentation tool based on requirements identified in an empirical analysis.

S. Voigt (✉)
Otto-von-Guericke-University/Fraunhofer IFF, Magdeburg, Germany
e-mail: stefan.voigt@iff.fraunhofer.de

D. Hüttemann · A. Gohr · M. Große
CosmoCode GmbH, Berlin, Germany
e-mail: huettemann@cosmocode.de

A. Gohr
e-mail: gohr@cosmocode.de

M. Große
e-mail: grosse@cosmocode.de

© Springer International Publishing AG 2018
Á. Rocha and L.P. Reis (eds.), *Developments and Advances in Intelligent Systems and Applications*, Studies in Computational Intelligence 718,
DOI 10.1007/978-3-319-58965-7_5

This paper is an extended and updated version of our CISTI2016 paper [34]. Since our work follows a design science research (DSR) approach [15], our paper is structured as follows [12]: This introduction is followed by a literature review (2). Section 3 presents our research method and Sect. 4 describes the concept of an agile documentation tool. Section 5 concludes this paper with a discussion and conclusion. Since this paper is part of an ongoing German research project, the evaluation section is brief (see Sect. 5.1).

2 Literature Review

2.1 Documentation in Agile Environments

The software industry played a dominant role in the development and dissemination of agile methods around the turn of the millennium [7], thus spawning great demand for research at the beginning of the twenty-first century. A sound overview of the state-of-the-art is provided in [3, 6, 9, 10]. While literature on documentation in agile environments also exists, it is confined to specific aspects of documentation.

One study presents the status quo of documentation strategies employed in agile projects [16]. Two studies analyze agile teams' documentation routines when delivering projects very similarly [27, 28]. Whereas these studies provide very useful insight into practice, they fail to provide any concrete solutions or theoretical foundations.

A closer look at artifacts used in agile projects suggests that every artifact can be regarded as documentation [13]. Hadar et al. focus on architecture documentation and develop an abstract architecture specification [14]. Kanwal et al. examine offshore agile projects and develop key documents needed to communicate with off-site team members [17]. The research question driving the development of an approach to reducing the documentation required by the software standard IEC 61508-3:2010 is how to use agile methods for safety critical software [19]. Another document-driven approach is furnished by [29] along with suggestions for requisite documents. Active Documentation Software Design [22], on the other hand, is an approach to modeling the domain knowledge needed in agile projects. Such knowledge should be integrated into the source code in a special domain knowledge representation layer. Although the groundwork has clearly been laid, a complete method of or tool for documentation in agile projects is still needed. That is why we have developed a method of agile documentation (the details of which exceed the scope of this paper) in our research project, which can be integrated with the tool concept presented in Sect. 4.

2.2 Agile Documentation Tools and Wikis

The earliest work on agile modeling [4] and agile documentation [23] employed simple tools such as note cards and whiteboards to record information temporarily. Wikis have been identified as the best electronic tool for this, though [4, 23]. A document management system is essential to the storage of the documents outlined in the documented approach [30] but it is not a suitable tool for tracing information.

A wiki's users develop and link its content collaboratively. Traditional wikis, however, do not provide (semantic) information representation capabilities essential to traceability and complex information structures (e.g. correlations between requirements and solutions). Different types of wikis, including semantic, structured and hybrid wikis, now exist, which enable users not only to manage unstructured information but also to add structures to them [32].

A wiki by itself is merely one tool that may have to interact with other systems during the development process, too. Other established tools such as integrated development environments (IDE), version control systems (VCS) and issue trackers are also utilized in agile development environments [1]. Information has to be traceable between the different tools [25].

Wikis are already being used in agile environments for documentation. A flexible wiki approach has been adopted for requirements documentation [25] but, as our study [33] reveals, requirements are already documented quite well (around 50% indicate that much or almost all information on requirements is documented). Concrete solutions and decisions on alternative solutions are documented far less (much or almost everything is documented by only 28% and 24%, respectively). Our research findings combined with the responses to our survey [33] led us to identify a need for the development of an agile documentation tool, which is based on wiki technology and can be integrated with other information sources.

3 Research Method

Excellent design science research (DSR) approaches are outlined in [31]. We chose to follow a DSR approach [20] that includes the following steps: problem identification and motivation, definition of goals, design and development, demonstration, and evaluation and communication [20]. Our "sprintDoc" research project consortium consists of one university, one software development company and two pilot users that also develop software. All of the project members use agile methods.

We applied method triangulation to obtain an overall picture of the needs and requirements in agile development. Our analysis was the outgrowth of four workshops (similar to project retrospectives), nine interviews with our pilot users, and an online survey of other software development companies (99 respondents) [33]. Although our analysis is not the focus of this paper, some of the results are introduced to underpin the argumentation behind our concept.

4 Concept of the "SprintDoc" Agile Documentation Tool

4.1 Overall Use Case, Structure and Interfaces

Since an agile documentation tool—we call our prototype sprintDoc—does not function as a stand-alone system, it has to be integrated in a company's agile system landscape. We examined this in-depth and determined that the dominant tools in the real-world IT landscape are[1]:

- Issue/bug trackers (including agile project management),
- Wikis,
- VCSs (including continuous integration systems), and
- IDEs [1].

These tools are intended for different contexts in the development process. Issue trackers and wikis are employed on the functional level, while IDEs and VCSs are applied to source code. VCSs and issue trackers are used to manage and track changes. IDEs and wikis are tools for software being produced.

Our interviews (n = 9) revealed that developers need concrete requirements and implemented (coded) solutions from past projects. They additionally reported that they obtain information by analyzing issues and code. This, in turn, revealed that source code and issues have to be interconnected for the purpose of documentation. The sprintDoc tool should be integrated in a specified landscape and respond flexibly to every different focus. We developed the following use case together with the pilot users (see Fig. 1) and set up our prototype using the DokuWiki[2] wiki engine: The main idea is to integrate the development of documentation artefacts (wiki pages) in the agile process and thus trace changes in documents along with changes in issues. The common practice of tagging a commit with the issue ID is transferred to wiki page creation and modification. Changes to the wiki are also tagged with the issue ID: A team member picks a ticket from the issue tracker[3] to develop a feature. The issue tracker is used to discuss and refine development tasks

---

[1]We identified this landscape among our pilot users and in our study. At least two thirds of all companies surveyed use the tools listed (92% IDE, 90% VCS, 80% issue trackers, 68% wikis). Our study's pilot user and respondents also communicated need for integration with issue trackers, VCS and IDE [33].

[2]CosmoCode developed the widely used DokuWiki (www.dokuwiki.org) wiki engine on which we based our prototype.

[3]We chose Atlassian JIRA for our prototype and for use in our pilot user environments because it has been widely adopted [29].

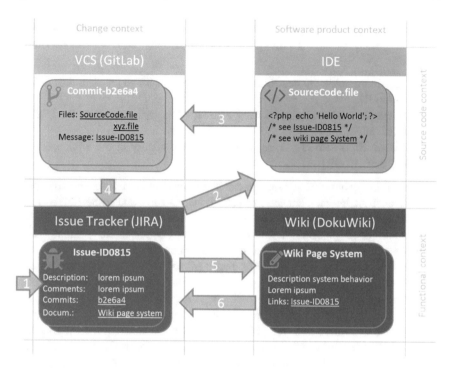

Fig. 1 Use case and integration in the tool landscape

in the comments on a ticket (1). The developer keeps the ticket ID while coding the feature in the IDE (2). When the developer commits the tested and optimized code to the VCS,[4] the commit message transfers the ticket ID to the VCS (3). The developer changes the issue's status to acceptance or documentation in the issue tracker (depending on the concrete process and the "definition of done") (4). The issue tracker displays links to the wiki to document the issue (5). Once system components affected by the issue have been documented in the wiki, the developer is linked with the issue tracker again to set the issue status to "done" (6).

The links between the JIRA issue tracker and DokuWiki are set automatically by the MagicMatcher plugin described below.

4.2 The MagicMatcher Function

Our interviews revealed that documentation is usually compiled immediately after or even parallel to (task-oriented) implementation. This made it clear that the tools have to interact very closely. Since the wiki should (automatically) know where

---

[4]The sprintDoc prototype is connected with the GitLab VCS for the same reasons.

system components, solutions or lessons learned have to be documented whenever work on certain issues is being completed, we included an issues session in the wiki:

- Special links guide the user from JIRA to certain wiki pages. The link delivers the issue ID.
- All changes made on pages during a session are connected with the issue ID automatically.
- The issue ID is also displayed in the document history of the wiki pages and linked back to JIRA.

The wiki is given a new issue header that is displayed in view and edit mode. The header contains the elements: project, issue ID and issue name (linked to the issue in JIRA) along with status and a suggestions page link (see Fig. 2). The suggestions page presents a list of suggested documentable content related the issue, which has been computed by the MagicMatcher.

The MagicMatcher stores issue IDs, pages edited during issue sessions, commits tagged with the issue ID and files contained in the commits (and thus connected with the issue ID). As the number of wiki pages grows, the MagicMatcher helps the user find the most fitting place to document information on the issue on which the developer is currently working. These suggestions are calculated on the basis of matches between source files that were altered during work on that issue and tagged with the issue ID during commit. A relation model consisting of "page ~ issue ~ commit ~ file" was developed as the basis for the MagicMatcher. The following relations and values were defined:

- CommitsIn (issue) is an issue's set of commits. A commit and an issue are affiliated with each other when the commit message contains the issue ID.
- FilesIn (commit) is a commit's set of source files. A file is affiliated with a commit when it is saved in the VCS in a commit.
- FilesIn (issue) is an issue's set of source files. A file is affiliated with an issue when it is saved by a commit and is affiliated with the issue. The weight of affiliation is determined by the number of source code lines modified (see (1)).

Fig. 2 Issue header in the DokuWiki (*top*)

$$(\text{weight}(\text{issue, file}) = \sum \text{modified lines} / \sum \text{lines}). \qquad (1)$$

- PagesDocumented (issue) is the set of pages documented in connection with an issue. A wiki page documents information affiliated with an issue when it is modified during an issue session.
- IssuesDocumented (pages) is the set of issues documented on a page. This is the inversion of PagesDocumented (issues).
- IssuesRelatedTo (issue) can be broken down into:

  - A relation based on issue properties. An issue has a relation to another issue when the two are linked.
  - A relation based on the congruence between commit and source files. An issue has a relation to another issue when the set of corresponding files is not empty. The weight of the relationship ensues from the weighting of affiliated files (see (2)).

$$\text{weight}(\text{issue1, issue2}) = \frac{\sum_{i=1}^{N}(\text{issue1, } f_i) \times \text{weight}(\text{issue2, } f_i)}{\sum_{i=1}^{N} \text{weight}(\text{issue2, } f_i) + \sum_{i=1}^{M} \text{weight}(\text{issue2, } f_i)} \qquad (2)$$

where N is the number of files in issue1 and M is the number of files in issue2.

The figure summarizes the given relations.

We assume that two issues are similar and connected when they share common source code files. In the given example (see Fig. 3), the files 1, 2 and 3 belong to commit1 and the files 2, 3, 4 and 5 to commit2. Both commits are therefore related to issue1 (see relation FilesIn (issue)).

A wiki page p is a candidate for documentation related to an issue i, when:

- the issue of p has a relation to i.
- p contains a link on i or,
- p of a component is assigned to i.

Figure 4 presents the results of PageCandidatesFor (issue) for issue "TEST-2".

Computations for our relation model are initiated by WebHooks, i.e. a WebHook triggered by pushing to the VCS calls a function in the MagicMatcher to start computation. The following events must be computed:

- Committing to the VCS.
- Moving or deleting source code files.
- Editing wiki pages during an issue session.
- Renaming, moving or deleting a wiki page.
- Marking certain source code files as irrelevant for the computation.

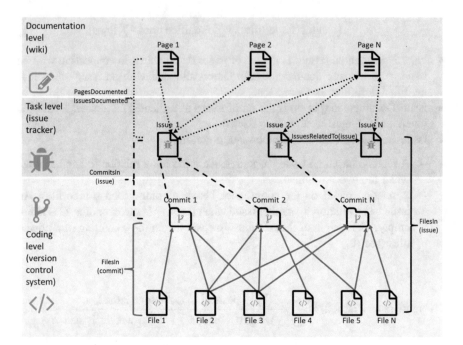

Fig. 3 Relation model calculated with the MagicMatcher

4.3   Struct and Farming Plugin

Two other important parts of our documentation tool deserve brief mention. The MagicMatcher helps identify the best location to document certain issues. Most documentation use cases need more and flexible structures.

Since the client is heavily involved in the entire software development process in agile projects, the client (e.g. the product owner in scrum projects) should be included in documentation processes as well. This necessitates upper level structures, i.e. parts of the documentation tool should be encapsulated in terms of content and rules. The client should not see any projects other than the one in which the client is involved. We found the farming concept to be very helpful. A new wiki ("animal") based on a blueprint ("farmer") should be easily setup for each client. The intention is to make the process as simple as possible in order to reduce administrative work. We developed the farmer plugin to develop and maintain the wiki farm (https://www.dokuwiki.org/plugin:farmer) and the farmsync plugin to share identical content between animals (https://www.dokuwiki.org/plugin:farmsync).

Different structures that document different types of information are needed on a deeper level. For instance, a software project consists of different meetings at which the client and developers discuss new requirements, acceptance of features or lessons learned in the last iteration. All of these types of information can be

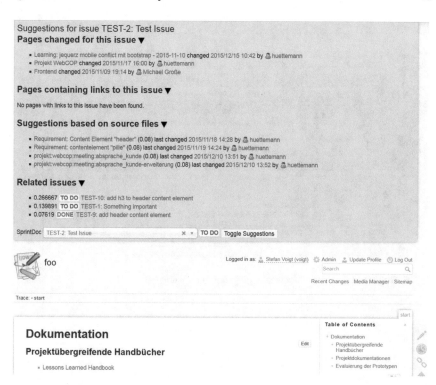

Fig. 4 Pages suggested for the issue "TEST-2" in DokuWiki

documented on different wiki pages, but automatically connecting related information would be more helpful (see Fig. 5). This is why we developed the struct plugin (https://www.dokuwiki.org/plugin:struct).

We use this plugin to attach structured data to wiki pages. This data is classified in schemas, a schema containing a set of fields with a specific type. These types control how data is displayed and validated against the input. This can be configured within the schema. Schemas and their assignments to certain pages (by namespace patterns) are managed centrally in an admin interface. Since the structured data are no longer part of the page syntax as in former concepts [32], a change in a schema changes it on every associated page. The structured data are only edited by means of a dedicated form in the standard editor or by inline editing aggregation.

Schemas, assignments and structured data are stored in a SQLite database. Different editors help users build and maintain schemas, assignments, assigned data or aggregations (see Fig. 6). These mechanisms enable us to attach data to a certain page to describe a requirement, a meeting or a lesson learned in a structured manner. A meeting can thus be connected with attending team members or certain requirements can be discussed by automatic aggregation.

Users can test the struct plugin with our demo system (see https://demo.sprintdoc.de/).

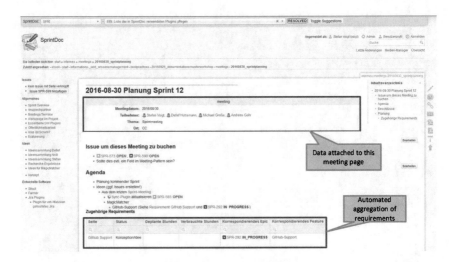

Fig. 5 Structured data and automatic aggregation on a wiki page

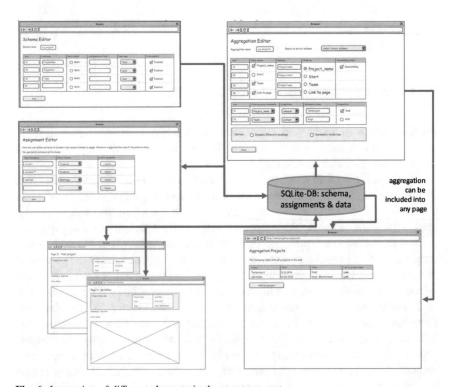

Fig. 6 Interaction of different elements in the struct concept

5 Evaluation and Conclusion

5.1 Evaluation

Since our paper is the outgrowth of an ongoing research project, we are unable to provide a final evaluation of our concept at this time. We are able to discuss initial findings, though. In a first step, we collected and analyzed feedback from the pilot user companies. Both of the pilot user companies were given customized prototypes accompanied by a survey evaluating usability, usefulness, fault tolerance or self-descriptiveness, etc. The pilot users recorded their feedback as they were working with the system. In a second step, we prioritized the change requests before programming changes. This enabled us to improve our prototype by adding the following new features:

- Setting or deleting a relation between an issue and a wiki page manually:

  - Connecting an existing page to a selected issue with one click.
  - Linking to an existing issue in JIRA by a link wizard.
  - Deleting a connection between a wiki page and an issue (including undo).

- Managing multiple projects in one wiki instance (animal).
- Preventing accidental connections of a wiki page and an issue during an issue session (two buttons "Save with issue X" and "Save without issue").

The first responses from pilot user companies have been positive and enabled us to implement the aforementioned improvements. The next steps of evaluation will entail thorough testing of the prototype by pilot users in real software development projects. Furthermore, we intend to use student groups to test whether our tool helps them write good documentation. Last but not least, we are eagerly awaiting feedback on the accuracy of the suggestions made by MagicMatcher.

5.2 Conclusion

Since information on software projects exists for up to three years, as over half of our survey respondents indicated, it is important to keep it accessible. Interviewees' identification of a lack of time as a chief argument against documentation indicates that documentation ought to be integrated in the tool landscape of the development process. Our online survey revealed that the three main requirements for documentation tools were that they keep documentation up-to-date, keep information traceable, and facilitate documentation quality assurance. Our prototype will therefore eliminate the chief problems of agile documentation, thus making it easier to keep documentation up-to-date, to find the correct location for storage, and to eliminate duplicate documentation. The prototype is currently being evaluated in our two pilot user companies. The initial feedback is promising.

Acknowledgements This study was funded by the German Federal Ministry of Education and Research's program "Innovative SMEs: ICT" overseen by the German Aerospace Center's Project Management Agency for Information Technology [01IS15005A-D].

References

1. 8th Annual State of Agile Development. http://www.versionone.com/pdf/2013-state-of-agile-survey.pdf (2014). Accessed 3 June 2015
2. Abrahamsson, P., Babar, M.A., Kruchten, P.: Agility and architecture: can they coexist? IEEE Softw. **27**, 16–22 (2010). doi:10.1109/MS.2010.36
3. Abrahamsson, P., Conboy, K., Wang, X.: Lots done, more to do': the current state of agile systems development research. Eur. J. Inf. Syst. **18**, 281–284 (2009)
4. Ambler, S.W.: Agile Modeling: Effective Practices for Extreme Programming and the Unified Process. Wiley, New York (2002)
5. Bouillon, E., Güldali, B., Herrmann, A., Keuler, T., Moldt, D., Riebisch, M., Bouillon, E., Güldali, B., Herrmann, A., Keuler, T., Moldt, D., Riebisch, M.: Leichtgewichtige Traceability im agilen Entwicklungsprozess am Beispiel von Scrum. Softwaretechnik-Trends **33**, 29–30 (2013). https://www.springerprofessional.de/leichtgewichtige-traceability-im-agilen-entwicklungsprozess-am-b/5048954, http://pi.informatik.uni-siegen.de/gi/stt/33_1/01_Fachgruppenberichte/RE_FG-Treffen/14_bouillon.pdf
6. Cohen, D., Lindvall, M., Costa, P.: Agile Software Development. New York (2003)
7. Conboy, K., Morgan, L.: Future research in agile systems development: applying open innovation principles within the agile organisation. In: Dingsøyr, T., Dybå, T., Moe, N.B. (eds.) Agile Software Development, pp. 223–235. Springer, Berlin, Heidelberg (2010)
8. Dingsøyr, T., Dybå, T., Moe, N.B.: Agile software development: an introduction and overview. In: Dingsøyr, T., Dybå, T., Moe, N.B. (eds.) Agile Software Development. Springer, Berlin, Heidelberg, pp. 1–13 (2010)
9. Dingsøyr, T., Nerur, S., Balijepally, V.G., Moe, N.B.: A decade of agile methodologies: towards explaining agile software development. J. Syst. Softw. **85**, 1213–1221 (2012). doi:10.1016/j.jss.2012.02.033
10. Dybå, T., Dingsøyr, T.: Empirical studies of agile software development: a systematic review. Inf. Softw. Technol. **50**, 833–859 (2008). doi:10.1016/j.infsof.2008.01.006
11. Glass, R.L.: Agile versus traditional: make love, not war! Cut. IT J. **14**, 12–18 (2001)
12. Gregor, S., Hevner, A.R.: Positioning and presenting design science research for maximum impact. MIS Quaterly **37**, 337–356 (2013)
13. Gröber, M.: Investigation of the Usage of Artifacts in Agile Methods. Master Thesis, Munich (2013)
14. Hadar, I., Sherman, S., Hadar, E., Harrison, J.J.: Less is more: architecture documentation for agile development. In: Prikladnicki, R., Hoda, R., Cataldo, M., Sharp, H., Dittrich, Y., de Souza, C. (eds.) 6th International Workshop on Cooperative and Human Aspects of Software Engineering (CHASE), pp. 121–124 (2013)
15. Hevner, A.R., March, S.T., Park, J., Ram, S.: Design science in information systems research. MIS Quaterly **28**, 75–105 (2004)
16. Hoda, R., Noble, J., Marshall, S.: Documentation strategies on agile software development projects. IJAESD **1**, 23 (2012). doi:10.1504/IJAESD.2012.048308
17. Kanwal, F., Bashir, K., Ali, A.H.: Documentation practices for offshore agile software development. Life Sci. J. **10**, 70–73 (2014)
18. Lethbridge, T.C., Singer, J., Forward, A.: How software engineers use documentation: the state of the practice. IEEE Softw. **20**, 35–39 (2003). doi:10.1109/MS.2003.1241364

19. Myklebust, T., Stålhaneb, T., Hanssena, G.K., Wienc, T., Haugseta, B.: Scrum, documentation and the IEC 61508-3:2010 software standard. In: Proceedings of the Probabilistic Safety Assessment and Management (2014)
20. Peffers, K., Tuunanen, T., Rothenberger, M.A., Chatterjee, S.: A design science research methodology for information systems research. J. Manag. Inf. Syst. **24**, 45–77 (2007). doi:10.2753/MIS0742-1222240302
21. Prause, C.R., Durdik, Z.: Architectural design and documentation: waste in agile development? In: Jeffery, R., Raffo, D., Armbrust, O., Huang, L. (eds.) Proceedings of the 2012 International Conference on Software and System Process (ICSSP), pp. 130–134 (2012)
22. Rubin, E., Rubin, H.: Supporting agile software development through active documentation. Requir. Eng. **16**, 117–132 (2011). doi:10.1007/s00766-010-0113-9
23. Rüping, A.: Agile Documentation: A Pattern Guide to Producing Lightweight Documents for Software Projects. Wiley, Hoboken, NJ (2003)
24. Selic, B.: Agile documentation, anyone? IEEE Softw. **26**, 11–12 (2009). doi:10.1109/MS.2009.167
25. Silveira, C., Faria, J.P., Aguiar, A., Vidal, R.: Wiki based requirements documentation of generic software products In: Proceedings of the 10th Australian Workshop on Requirements Engineering (AWRE'2005), pp. 42–51 (2005)
26. Stettina, C.J., Heijstek, W.: Necessary and neglected?: an empirical study of internal documentation in agile software development teams. In: Protopsaltis, A., Spyratos, N., Costa, C.J., Meghini, C. (eds.) Proceedings of the 29th ACM international conference on Design of communication, pp. 159–166 (2011)
27. Stettina, C.J., Heijstek, W., Faegri, T.E.: Documentation work in agile teams: the role of documentation formalism in achieving a sustainable practice. In: IEEE Computer Society (ed.) Proceedings-2012 Agile Conference, pp. 31–40 (2012)
28. Stettina, C.J., Kroon, E.: Is there an agile handover? an empirical study of documentation and project handover practices across agile software teams. In: 2013 International Conference on Engineering, Technology and Innovation (ICE) & IEEE International Technology Management Conference (2013)
29. Tripathi, V., Goyal, A.K.: A document driven approach for agile software development. IJARCSSE **4**, 1085–1090 (2014)
30. Uikey, N., Suman, U., Ramani, A.K.: A documented approach in agile software development. IJSE **2**, 13–22 (2011)
31. Vaishnavi, V., Kuechler, B.: Design science research in information systems. http://desrist.org/desrist/content/design-science-research-in-information-systems.pdf (2004). Accessed 17 July 2015
32. Voigt, S., Fuchs-Kittowski, F., Gohr, A.: Structured wikis: application oriented use cases. In: Proceedings of the 10th International Symposium on Open Collaboration. ACM, New York (2014)
33. Voigt, S., von Garrel, J., Müller, J., Wirth, D.: A study of documentation in agile software projects. In: Proceedings of the 10th ACM/IEEE International Symposium on Empirical Software Engineering and Measurement. ACM, New York (2016)
34. Voigt, S., Hüttemann, D., Gohr, A.: sprintDoc: concept for an agile documentation tool. In: Rocha, A., Reis, L.P., Cota, M.P., Suárez, O.S., Gonçalves, R. (eds.) Sistemas y Tecnologías de Información, pp. 1146–1149 (2016)

Pervasive Business Intelligence: A Key Success Factor for Business

Teresa Guarda, Marcelo León, Maria Fernanda Augusto,
Filipe Mota Pinto, Oscar Barrionuevo and Datzania Villao

Abstract Today the strategic significance of information is fundamental to any organization. With the intensification of competition between companies in open markets and often saturated, companies must learn to know themselves and to the market through the collection and analysis of quality information. The strategic information is seen as a key resource for success in the business, which is provided by Business Intelligence systems. A successful business strategy requires an awareness of the surrounding (internal and external) environment of organizations, including customers, competitors, industry structure and competitive forces. Managing the future means not only is able to anticipate what will happen outside the organization, but also be able to represent the events through their own actions timely. To make it possible, Pervasive Business Intelligence arises as a natural evolution of business intelligence applications in organizations, allowing to companies achieve and maintain a sustainable competitive advantage, helping managers react proactively in a timely manner to threats and opportunities.

1 Introduction

Increasingly there are a greater number of organizations that provide Business Intelligence (BI) to their decision makers (internal and external). Internally, reinforces the responsibility of all the collaborators and the improve management

T. Guarda (✉) · O. Barrionuevo
Universidad de las Fuerzas Armadas—ESPE, Sangolqui, Quito, Ecuador
e-mail: tguarda@gmail.com

T. Guarda
Algoritmi Centre, Minho University, Guimarães, Portugal

F.M. Pinto
Instituto Politécnico de Leiria—ESTG, Leiria, Portugal

M. León · M.F. Augusto · D. Villao
Universidad Estatal Peninsula de Santa Elena—UPSE, Santa Elena, Ecuador

© Springer International Publishing AG 2018
Á. Rocha and L.P. Reis (eds.), *Developments and Advances in Intelligent Systems and Applications*, Studies in Computational Intelligence 718,
DOI 10.1007/978-3-319-58965-7_6

stability. Externally, relations with suppliers and business partners can be strengthened through effective sharing of performance indicators for mutual benefits [1].

It is very important and also difficult for organizations to make the right decisions. Companies know that the ability to make the right decisions is often essential for increased profits, for risk management and for good overall performance. Due to uncontrollable factors such as the fast-moving markets, the economic and regulatory changes, and new sources of competition, making the right decision is not a peaceful issue.

BI can be understood as the use of different sources of information to define the competitive strategies of an organization [2]. BI goes from the process of collecting large amounts of data, its analysis, and consequent production of reports that summarize the essence of actions on the business, which will assist the managers in the decision making of the day-to-day business [3, 4]. Thus, we can consider that BI is the process through which users obtain accurate and consistent data from the storage of organizational data environment. The data obtained from various business contexts, allow users to identify, analyze and detect trends, opportunities, threats and anomalies, and make predictions. BI systems and tools play a key role in organizational strategic planning process. These systems allow collect, store, access and analyze data in order to support and facilitate decision making process [5]. The organizations develop their strategies to maintain or achieve a sustainable competitive advantage, thus being hostages of BI systems and tools.

Pervasive Business Intelligence (PBI) is a management concept that refers to a collection of tools and technologies that provide capabilities to collect analyze and process data the organization data. Regardless of the size of the organization, the main objective of PBI is to assist in decision-making process, all levels of the organization timely.

The intensification of competition between organizations in nowadays saturated markets, makes organizations hostages and dependent of the information with strategic significance, which includes know themselves, the stakeholders and to the market. A successful business strategy requires an awareness of the internal and external environment of organizations, including their customers, competitors, industry structure and competitive forces.

This chapter presents a framework for Pervasive Business Intelligence as a key factor to enable organizations to gain or maintain a sustainable competitive advantage. The chapter is organized as follows. In this introductory section is dedicated to the presentation. The 2nd section discusses PBI. The next section (3rd) is dedicated to presents the propose PBI framework for achieving competitive advantage. In 4th section, are presented the final remarks and future research options.

2 Pervasive Business Intelligence

The ability to make the right decisions timely is essential for increased organization profits and a good performance. Companies are emerged in an environment of uncontrollable factors: economics, regulatory, markets, competition and others; and decide correctly at the right time is not a peaceful task.

Based on existing studies, we found three different approaches to BI: a management approach, a technical approach and a value-added approach. The management approach addresses the BI as a integration process of data collected from the organizational environment (internal and external) in order to be able to extract the relevant knowledge for management decision making [6–9]. In the case of technical approach, BI is presented as a set of tools that support the process outlined by the management approach. The emphasis is given to the technology used, not to the process itself [10–12]. In the value-added approach, BI systems provide added value in the acquisition of competitive advantage [13–15].

In the current situation, the markets are mature and saturated, and exposed to fierce competition; companies are forced to seek alternative ways to increase the value of their BI initiatives, being greater the effort to achieve PBI [16]. The focus will be disseminate BI across all areas of the business, and BI systems become part of business processes, with flexibility to adapt business changes and information needs [17].

There are various definitions of PBI, is the ability to deliver timely manner to all users, the integrated information in data warehouses (DW), providing the necessary visibility, knowledge, and facts for decision making in all business processes [18]; is the improvement of the capabilities of making strategic and operational decision of an organization, through the design and implementation of the organizational culture of business processes and technologies as a whole [19]; is BI across the organization, providing to all people, and at all levels of the organization the analyzes, alerts and feedback mechanisms [20].

The implementation of PBI in organizations is supported by applications that access the data in real time, supporting the actions of supply chain management (SCM), and the actions of customer's relationship management (CRM). The application of PBI is increased when the employees are on the front line contact with customers and can create new sales opportunities, up-sell and cross-sell [18]. The PBI aims to align all processes, to allow the delivery of relevant information to users who need support in decision making.

There are five key factors with great influence in the dissemination of BI [19]: quality of the BI project, level of training, prominence of regulation, non-executive involvement, and use a methodology of performance assessment (Fig. 1). In the first key factor, the quality of the BI project, the expectations of users for the components of BI solutions are satisfied. The next key factors, the level of training, the user's degree of satisfaction with the training in the use of BI tools and analytic techniques to improve the decision making is high. The third key factor, prominence of regulation, it focuses on the importance of regulation and policies regarding data in BI systems. The key factor non-executive involvement represents the involvement of non-executives in disseminating and promoting of the use of BI tools in the organization. The last key factor, using a methodology of performance assessment is focuses in the importance of using a formal methodology for assessing performance within the organization.

A company with BI systems can integrate powerful tools, monitoring system with various metrics, data integration, among other features, analysis, standardized reporting within a service-oriented architecture [21], and that is essential for a good business management, guiding managers for quality information, with the establishment of standards and procedures to ensure compliance with the objectives [22]. It is not guaranteed that a BI system will generate a return on investment, but an adequate and integrated BI, can create the competitive advantage necessary for organizations. Once identified business processes, must be identified key stakeholders, the roles, the system functional requirements, the information needed for reporting, analysis, and presentation delivery should be defined. Companies that have adopted BI systems can compete more effectively in the marketplace, with additional information about their customers, suppliers, concurrence as well as a more efficient financial management. Then, companies should adopt a strategic and active behavior, adjusting strategies and becoming more competitive compared to competitors [23].

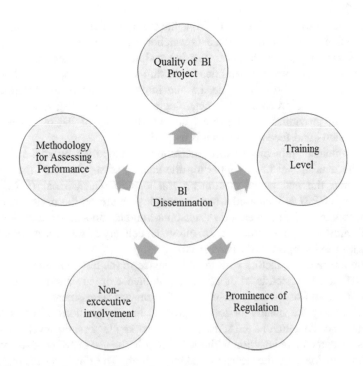

Fig. 1 Key success factors in BI dissemination

PBI arises as a normal evolution of BI applications in organizational scope, with application from the strategic level to operational level, and appears on the last level (5th) of maturity of BI, the pervasive level. The 1st level of maturity, is characterized by inconsistent data, with a high use of spreadsheets and limited use of adequate communication tools, users have high dependence on IT department. In the 2nd level, it can be seen some investment in BI, but the majority of are not sufficiently qualified to take advantage of the system, and managers do not trust in information provided. At the 3th level, organization has the first success and therefore the first benefits to the business, although, this level is characterized by the absence of application data integration. In the 4th maturity level, the organizational already have a defined strategy for develop BI; the strategic information is trustable and is used to support strategic decision making process. Users have the appropriate knowledge to use BI tools, and data quality is supervised constantly. In the last level (5th), BI is disseminated across all organization, crossing all business areas of the business with the organizational culture. BI systems are integrated into the business processes, enabling that organizational changes be adapted to meet the information needs,

PBI is the operationalization of BI throughout the organization enabling BI systems reaches all levels of the organization, at the right time and with the necessary information.

3 Achieve Competitive Advantage

We can understand competitive advantage as a core competence that will give to the organization a strong competitive position compared with the competition. An organizations achieves competitive advantage when offer a superior value to customers. The efforts made in search for a competitive advantage are consolidated in the company strategy [24].

Advanced information of the variables present in a scenario of a competitive environment has a high strategic value. If a company is able to anticipate favorable/unfavorable situations for a particular scenario, then is possible develop an appropriate strategy for action. In this sense, investments in competitive intelligence are always very opportune [25].

Companies need to develop their own strategies to gain competitive advantage over its competitors. Any successful company has one or two functions it performs better than the competition (core competence). If a core competence of a company offers an advantage in the market over the long term, it is called the persistent competitive advantage. For a core competency reach this level will have to become difficult to imitation, single, persistent, higher than the competition, and applicable to many situations.

In today's world with a rapidly adaptive competition, none of these advantages can persist for a long term. The only way of competitive advantage to be truly sustainable is to build an organization so alert and agile that can always detect benefits, opportunities, and threats immediately, regardless of changes in the market, being fundamental the support provided by PBI systems.

We can see BI systems as a set of decision support systems (DSS) that allows decision makers to direct their actions according to the organization strategy. For DSS be successful, decision making process is critically dependent upon the availability high quality information integrated, organized and presented timely [26, 27].

To create and maintain competitive advantage, companies need to adapt constantly, changing business processes to meet the needs and expectations of customers, suppliers, stakeholders and changes in the business environment (Fig. 2). To improve business processes, companies have to make the necessary changes, being necessary to redesign processes. In this context, managers need a method that allows them to determine timely a concise manner to do that, the processes that no longer meet the needs of the business and need to be redesigned. Enterprise systems support business processes, recording the nature of the operations that are

Fig. 2 Pervasive business intelligence framework

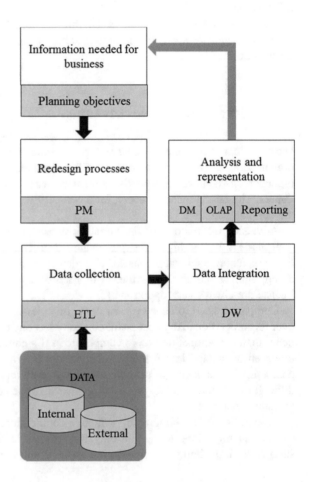

performed, being possible to build and understand the real models of existing business processes. Process Mining (PM) refers to the tools and techniques that allow the extraction of knowledge of available event logs in enterprise information systems. These techniques and tools provide new ways to discover, monitor and improve processes. The process mining assumes that it is possible to obtain the flow of activities for a process from execution logs of transactions made in information systems. The initial process mining techniques, allow achieving satisfactory results for well-structured processes, but failed in the case of unstructured or poorly defined processes, which lacked a strong dependency between activities [28]. The processes mapping is vital for companies who wish to align their processes with business strategy activity. The mapping of processes of a company, allows faithfully retract information flows, identify their weaknesses, existing inconsistencies, which support the flow of information (digital, physical). The objective of process mining is the extraction of information from the logs in order to capture the business process, in the way it was executed.

The PBI include technologies and tools. The technologies include data warehousing (DW), and on-line analytical processing (OLAP). A DW is a relational database that is designed for query and analysis, is a repository for all the data that various organization business systems collect. The OLAP processing, deals with the ability to analyze large volumes of information in various perspectives within a DW. The OLAP also refers to the analytical tools used in BI for visualization of management information and supports the functions of the organizational business analysis. The tools include data mining (DM) and PM. DM is a set of methods and techniques for sorting through data to identify patterns and make relationships between the data. PM is a set of techniques and tools that allow the extraction of knowledge from the logs of events available in organizational the information systems. PM purpose is to log the information extraction to capture the business process manner that it runs. PBI technology and tools help end-users in decision make, providing accurate, current and timely information.

4 Conclusion

The today's business environment and its complexity forces companies to be agile and proactive in relation to decision making processes [29], it is necessary to understand the information to track the history of sustainable future events, leading many organizations to adopt BI systems in its business processes [30]. Then, one of the keys of business strategy for creating competitive advantages is based on the understanding of the data that companies generate in its own business, and the information processing has gradually become the foundation for achieving competitive advantage, and organization has to believe that have the information needed at the right time [31]. BI systems and tools have a crucial role in decision making process, allowing collect, store, access and analyze organizational data in order to support and facilitate decision making [5]. BI tools have a number of advantages for businesses, emphasizing the reduction of the dispersion of information; improved access to information; real time information availability; flexibility and versatility in adapting to the reality of the company and usability useful in the decision making process [32].

PBI emerges from a natural evolution of the BI systems, with an application from the strategic level to the operational level. According Vesset, PBI is the improvement of the strategic and operational decision making capabilities in a organization through the design and implementation of it as a whole, including organizational culture, business processes, and technologies [19]. PBI aims to integrate and align all processes, to enable the delivery of relevant information which assists users in decision making process.

PBI allows decision makers to react in time, to the threats, problems, opportunities, supporting the creation/maintaining competitive advantage. The only truly sustainable competitive advantage is to build an organization so alert and agile that can always detect benefits, opportunities, and threats immediately, regardless of changes in the market, being fundamental the support provided by PBI systems and tools [33].

Today's, the complexity of internal and external business environment increases the need for pervasive business intelligence. PBI will be achieved only when the BI is integrated into the organization's business processes, and being an integral part of the decision making process too. Just in these circumstances, pervasive business intelligence systems could help managers react in a timely manner to threats and opportunities, being proactive and reactive [34].

In future research options our intention is to improve the framework and test it with ontology models.

References

1. Xie, G., Yang, Y., Liu, S., Qiu, Z., Pan, Y., Zhou, X.: EIAW: towards a business-friendly data warehouse using semantic web technologies. Semant. Web 857–870 (2007)
2. Barbieri, C.: Business Intelligence: Modelagem & Tecnologia. Axcel Books (2001)
3. Stackowiak, R., Greenwald, R.: Oracle Data Warehousing and Business Intelligence Solutions. Wiley Publishing, Indianapolis (2007)
4. Palocsay, S.W., Markham, I.S., Markham, S.E.: Utilizing and teaching data tools in excel for exploratory analysis. J. Bus. Res. 191–206 (2010)
5. Aaker, D., Kumar, V., Day, G., Leone, R.: Marketing Research, 10th edn. Wiley (2009)
6. Bucher, T., Gericke, A., Sigg, S.: Process-centric business intelligence. Bus. Process Manag. J. 408–429 (2009)
7. Cheng, H., Lu, Y.C., Sheu, C.: An ontology-based business intelligence application in a financial knowledge management system. Expert Syst. Appl. 3614–3622 (2009)
8. Bose, R.: Advanced analytics: opportunities and challenges. Ind. Manag. Data Syst. 155–172 (2009)
9. Lim, A.: Processing online analytics with classification and association rule mining. Knowl. Based Syst. 23(3), 248–255 (2010)
10. Baars, H., Kemper, H.G.: Management support with structured and unstructured data: an integrated business intelligence framework. Inf. Syst. Manag. 132–148 (2008)

11. Sahay, B., Ranjan, J.: Real time business intelligence in supply chain analytics. Inf. Manag. Comput. Secur. 28–48 (2008)
12. Chen, M.K., Wang, S.C.: The use of a hybrid fuzzy-Delphi-AHP approach to develop global business intelligence for information service firms. Expert Syst. Appl. 7394–7407 (2010)
13. Wang, H., Wang, S.: A knowledge management approach to data mining process for business intelligence. Ind. Manag. Data Syst. 622–634 (2008)
14. Fleisher, C.S.: Using open source data in developing competitive and marketing intelligence. Eur. J. Mark. 852–866 (2008)
15. Davenport, T.H., Harris, J.G.: Competing on Analytics. Harvard Business School (2007)
16. Ortiz, S.: Taking business intelligence to the masses. Computer 43, 12–15 (2010)
17. Rayner, N., Schlegel, K.: Maturity Model Overview for Business Intelligence and Performance Management. Gartner Inc. Research (2008)
18. Markarian, J., Brobst, S., Bedell, J.: Critical Success Factors Deploying Pervasive BI. Informatica, Teradata, MicroStrategy (2008)
19. Vesset, D., McDonough, B.: Improving Organizational Performance Management Through. IDC (2009)
20. Mittlender, D.: Pervasive business intelligence: enhancing key performance indicators. Inf. Manag. 15(4), 11 (2005)
21. Eckerson, W.: Performance Dashboards: Measuring, Monitoring, and Managing Your Business. Wiley (2010)
22. Ranjan, J.: Business justification with business intelligence. VINE 461–475 (2008)
23. Reeves, M., Deimler, M.: Strategies for winning in the current and post-recession environment. Strateg. Leadersh. 37, 10–17 (2009)
24. Potter, M., Paulino, L.: Estratégias. Circulo de Leitores (2014)
25. Porter, M.: Competitive Advantage: Creating and Sustaining Superior Performance. Free Press, New York (1985)
26. Santos, M.F., Portela, F., Vilas-Boas, M.J., Abelha, A., Neves, J., Silva, A., Rua, F.: A pervasive approach to a real-time intelligent decision support system in intensive medicine. In: CCIS—Communications in Computer and Information Science, vol. 272, pp. 368–381. Springer (2012)
27. Chen, H., Chiang, R.H., Storey, V.C.: Business intelligence and analytics: from big data to big impact. MIS Q. 36 (4) (2012)
28. van der Aalst, W.M.P., Günther, C.: Finding structure in unstructured processes: the case for process mining. In: 7th International Conference on Application of Concurrency to System Design 2007 (ACSD 2007), pp. 3–12, Washington (2007)
29. Bocij, P., Greasley, A., Hickie, S.: Business Information Systems: Technology, Development and Management. FT Press (2009)
30. Marjanovic, O.: The next stage of operational business intelligence: creating new challenges for business process management. In: 40th Annual Hawaii International Conference on System Sciences 2007 (HICSS 2007), p. 215 (2007)
31. Palmer, A.: The Business and Marketing Environment. McGraw-Hill, London (2000)
32. Lönnqvist, A., Pirttimäki, V.: The measurement of business intelligence. Inf. Syst. Manag. 32–40 (2006)
33. Guarda, T., Pinto, F., Cordova, J., Mato, F., Quiña, G., Augusto, M.: Pervasive business intelligence as a competitive advantage. In: IEEE (ed.) 2016 11th Iberian Conference on Information Systems and Technologies (CISTI), pp. 1–4, June 2016
34. Guarda, T., Santos, M., Pinto, F.: Pervasive Business intelligence: a framework proposal. In: International Conference on Computer Science and Information Engineering, pp. 127–131. DEStech Publications, Bangkok, Thailand (2015)

Annotated Documents and Expanded CIDOC-CRM Ontology in the Automatic Construction of a Virtual Museum

Cristiana Araújo, Ricardo G. Martini, Pedro Rangel Henriques
and José João Almeida

Abstract The Museum of the Person (Museu da Pessoa, MP) is a virtual museum
with the purpose of exhibit life stories of common people. Its assets are composed
of several interviews involving people whose stories we want to perpetuate. So the
museum holds an heterogeneous collection of XML (eXtensible Markup Language)
documents that constitute the working repository. The main idea is to extract automatically the information included in the repository in order to build the virtual
museum's exhibition rooms. The goal of this paper is to describe an architectural
approach to build a system that will create the virtual rooms from the XML repository to enable visitors to lookup individual life stories and also inter-cross information among them. We adopted the standard for museum ontologies CIDOC-CRM
(CIDOC Conceptual Reference Model) refined with FOAF (Friend of a Friend) and
DBpedia ontologies to represent OntoMP. That ontology is intended to allow a conceptual navigation over the available information. The approach here discussed is
based on a TripleStore and uses SPARQL (SPARQL Protocol and RDF Query Language) to extract the information. Aiming at the extraction of meaningful information, we built a text filter that converts the interviews into a RDF triples file that
reflects the assets described by the ontology.

C. Araújo (✉) · R.G. Martini · P.R. Henriques · J.J. Almeida
Department of Informatics, Algoritmi Research Centre, University of Minho,
4710-057 Gualtar, Braga, Portugal
e-mail: decristianaaraujo@hotmail.com

R.G. Martini
e-mail: rgm@algoritmi.uminho.pt

P.R. Henriques
e-mail: prh@di.uminho.pt

J.J. Almeida
e-mail: jj@di.uminho.pt

© Springer International Publishing AG 2018   91
Á. Rocha and L.P. Reis (eds.), *Developments and Advances in Intelligent
Systems and Applications*, Studies in Computational Intelligence 718,
DOI 10.1007/978-3-319-58965-7_7

1 Introduction

The society is more and more concerned with the preservation and the dissemination of Cultural Heritage, as works of art, ancient objects, and documents, among others.

Nowadays this can be achieved in a better way resorting to the information and communication technologies because they allow that the physical objects, on one hand, become accessible to anyone, and on the other hand, are not deteriorated rectos [1–3].

In this context of technological expansion, increasing the capability of extraction, storage and visualization of everyday life events, the museums have taken advantage to expand its field of action, as well as their own concept. They expand their geographical borders by providing information in their pages on the Internet and exhibiting their collections. On the other hand, completely virtual environments (called Virtual Museums, VM) appeared, without any references to physical spaces [2].

A Virtual Museum, such as a traditional museum, also acquires, conserves, and exhibits the heritage of humanity (in that case, intangible objects, or immaterial things[1]) creating a delightful environment for pleasure or enjoyment, as well as an appropriate place for teaching, and research.

This article is concerned with the creation of a specific Virtual Museum, the Museum of the Person (MP). The assets of the MP contains several interviews that narrate the life stories of ordinary citizens. These citizens, to report their life stories, remember events and other particular situations they have participated in. MP resources are constituted by a collection of documents in XML (eXtensible Markup Language) format.

In the article we discuss the interest and the way of building a virtual museum (that we see was a virtual learning space) to tell to the world those life stories and to extract knowledge about an epoch and a society connecting and relating them.

More precisely we aim at rebuilding npMP, the Portuguese branch of the Museum of the Person network (this network includes branches in Brazil, Portugal, USA, Canada, etc.) that connect individuals and groups through sharing their life stories (http://www.museumoftheperson.org/about/).

In this paper, and after a brief introduction to MP (Sect. 1.1), we discuss the ontology built to describe the museums knowledge repository (Sect. 2), then we present different technical approaches to implement the desired virtual museum (Sect. 3) and, finally, we introduce and describe the first module of our system that extracts required information from XML repository and its storage in the triple store that instantiate the ontology (Sect. 4).

Besides OntoMP, an ontology for the museum of the person that is new and a first contribution of this work, also the extension of the standard CIDOC-CRM for museums with FOAF and DBpedia concepts and properties is another contribution presented. The discussion on DBpedia inclusion is new material not yet presented in previous conference version of this article.

---

[1] According to: http://www.unesco.org/culture/ich/index.php?lg=en\Źpg=00022#art2.

An important contribution of our work presented in the paper is the detailed definition of a generic architecture for the implementation of a system that creates the museum exhibition rooms from the documents repository. Moreover we designed and propose two possible implementations of that generic architecture, one more appropriate for situations where the repository is stored in a relational database, and the other to be used when the repository is archived in a triple-store. Our aim is to compare both approaches to understand the development effort involved in each one and to learn their benefits and drawbacks.

At the best of our knowledge there are not similar projects that use ontologies and tools to generate automatically virtual learning spaces from their specifications, neither in the scope of MP nor in the context of other virtual museums. So we will not include a section on related work. For the sake of space (necessary to introduce all the novelties of this paper) we decided not to include a state of the art section; the reader is referred to the authors pre-thesis [4, 5], where we review the form main topics: Ontologies and CIDOC-CRM; Cultural Heritage; Learning Spaces; and Virtual Museums.

1.1 Museum of the Person, an Overview

Museum of the Person aims at gathering testimonials from every human being, famous or anonymous, to perpetuate his history [1, 3].

Life stories are evidences in support of facts or statements attested by common people carrying a social and historical character, which must be preserved and processed to become an immeasurable human heritage (intangible or immaterial things). The interviewed are used as informers, reporting the events and emotions they experienced [1].

To report their life stories during a predefined structured interview, the narrators remember events and other particular situations they have participated in. These memories will act as a basic element for social research [1].

The Museum of the Person's collection consists of sets of XML documents, specified by a DTD (Document Type Definition created specially for that purpose and called MP-DTD) related with each participant. Typically each interview is split into three parts [6]:

- **BI:** a brief biography and personal data, such as name, date and place of birth, and job;
- **interview:** two versions of the interview are built and saved—the *interview* file refers to the raw interview and contains all the questions asked and the narrator's answers; the *edited* file is a plain text, structured by themes that define small portions of a person's life story. In this format, a life story may give rise to thematic stories (e.g., dating, childhood, craft, among others). Both *interview* and *edited* files contain metadata tagging;
- **photographs and their caption**. This caption includes a description of the image, people depicted, place and the date.

Aside the interviews, there is also a *thesaurus* that includes key concepts mentioned in the stories.

Details about the elements that constitute each DTD will be mentioned in the next section that will discuss the development of MPs ontology (OntoMP). For more details on Museum of the Person please see [7].

2   The CIDOC-CRM Ontology for MP, OntoMP

2.1   OntoMP: Original Design

After an exhaustive analysis of all the documents (XML instances, respective DTD's, and the thesaurus) that belong to Museum of the Person, we could identify the concepts and relations involved in the life stories. This first step enabled us to design OntoMP, an ontology for the Museum of the Person. In this way, the museum visitor can have a conceptual navigation over the collection.

The main concepts extracted from the analysis phase are: people *(pessoa)*, ancestry *(ascendência)*, offspring*(descendência)*, job *(profissão)*, house episode *(episódio casa)*, education episode *(episódio educação)*, dating episode *(episódio namoro)*, general episode *(episódio geral)*, childhood episode *(episódio infância)*, leisure episode *(episódio lazer)*, religious episode *(episódio religioso)*, accident *(evento acidente)*, migration *(migração)*, life's philosophy *(filosofia de vida)*, festivity *(festividade)*, catastrophic event *(evento catastrófico)*, political event *(evento político)*, marriage *(casamento)*, birth *(nascimento)*, dream *(sonho)*, uses *(costumes)*, religion *(religião)* [7, 8].

In a similar way we also identified the following relations: performs *(exerce)*, depicted *(éRetratada)*, visits *(visita)*, lives *(vive)*, receives *(recebe)*, tells *(narra)*, has *(tem)*, has-type *(tipo)*, enrolls *(participa)*, occurs *(ocorre)*, refers to *(dizRespeito)* [7, 8].

Then we realized that some more elements should be added to the ontology. The concepts added were: marital status *(estadoCívil)*, spouse *(cônjuge)*, widowhood *(viuvez)*, sex *(sexo)*, literacy *(habilitações literárias)*, political party *(partido político)*, first communion *(primeira comunhão)*, death *(morte)*, baptism *(batismo)*, child's birth *(nascimento do filho)*, photos *(fotos)*, description *(descrição)* and file *(ficheiro)* [7, 8].

The ontology so far obtained is depicted in Fig. 1.

Figure 1 shows the main concepts in a life story (ellipsis) related with Person and also shows his main data properties (rectangles). Figure 1 enhances Event concept (a relevant component of OntoMP) and its different sorts (subclasses).

To validate the ontology designed, we created some instances using actual life stories picked-up from the MP collection, as can be consulted in the projects site at the http://npmp.epl.di.uminho.pt. Notice that all those interviews were conducted in the past and we got written permissions to publish them.

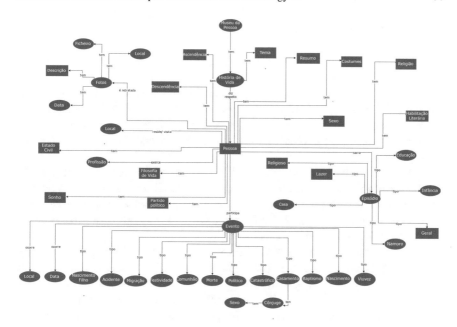

Fig. 1 An ontology *OntoMP: Original Design* for MP

2.2 OntoMP: CIDOC-CRM/FOAF/DBpedia Representation

After the validation and tuning of OntoMP, the next stage was to describe it in a standard ontology format used for museums, CIDOC-CRM (CIDOC Conceptual Reference Model). For that purpose we have followed the approach adopted in the context of another project to build Portuguese Emigration virtual museum [9].

CIDOC-CRM is a formal ontology planned to aid in the integration, mediation, and interchange of heterogeneous Cultural Heritage information [10]. It specifies the semantics of museums documentation.

CIDOC-CRM is an Event-based ontology, and therefore it should contain *Time-Spans* and *Places* related with each event. The core of CIDOC-CRM is based on seven concepts: *Temporal Entities, Events, Actors, Time-Spans, Conceptual Objects, Physical Things*, and *Places*. Notice that, *Actors* and *Conceptual Objects* or *Physical Things* should also be related with *Event* [10].

The transformation of *OntoMP: Original Design* in CIDOC-CRM was a straightforward process; the original concepts were expressed as events and associated concepts, and the original relations were mapped into the correspondent in CIDOC-CRM.

However, we found that some properties related with person could not be expressed in CIDOC-CRM in a simple and natural manner. So we decided to explore the combination with FOAF (Friend of a Friend) and DBpedia, since both contain a vocabulary specific to describe individuals, their activities and their relations with other people and objects [11].

FOAF ontology describes two areas of digital identity information: biographical and social network information [12].

DBpedia ontology is a shallow, cross-domain ontology, which has been manually created based on the most commonly used infoboxes within Wikipedia. DBpedia knowledge base covers various fields, such as geographic information, people, businesses, online communities, movies, music, books and scientific publications, among others [13].

After this investigation, we refined CIDOC-CRM adding some pertinent FOAF and DBpedia concepts and properties. Regarding FOAF, we imported *gender* property, person names (*name*, *givenName*, *familyName* and *nick*) and person-image relations (*depicts* and *depiction*). From DBpedia we picked up properties like *religion*, *profession*, *education*, *party* and *spouse*.

After the refinement of CIDOC-CRM ontology with FOAF and DBpedia elements, we got a simpler notation (descriptions became less verbose); moreover the original was enriched conceptually, this is more details about person's stories can be included in the knowledge base. The final OntoMP represented in this new notation was once again instantiated with concrete data extracted from the real life stories. It was possible to validate it once more.

In Fig. 2 we show an instance of the ontology created with data extracted from Maria Cacheira interview. Below we describe the CIDOC-CRM, FOAF and DBpedia fragment reproduced.

A person *(E21 Person)*, *gender* Female, *name* Maria Alice Rodrigues Cacheira (decomposed in *givenName* Maria Alice and *familyName* Rodrigues Cacheira), *participated in (E5 Event)* that is her birth *(E67 Birth)*. This event occurred at a *(E52 Time Span)*—that *is identified by (P78)* 1946-10-08, an *(E50 Date)*—and at a *(E53 Place)*—that *is identified by (P87)* Afurada an *(E44 Place Appellation)*.

This person *(E21 Person)* is *depicted* in the photo *(E38 Image)*. This photo *is identified by (P1)* 090-F-01.jpg *(E41 Apellation)*, *has note (P3)* Maria Alice Rodrigues Cacheira, *refers to (P67)* Maria Alice Rodrigues Cacheira *(E55 Type Description)*, and was taken in a *(E52 Time Span)*—that *is identified by (P78)* 2001-12-07, an *(E50 Date)*—and at a *(E53 Place)*—that *is identified by (P87)* Junta de Freguesia da Afurada, an *(E44 Place Appellation)*.

A person *(E21 Person)* has *education* "Sabe ler e escrever (4ª classe)", *professes* the *religion* "Católica" and has *profession* "Peixeira e Empregada de limpeza".

In this fragment of Maria Cacheira's life story other concepts can be identified. All these concepts, that characterize a *(E21 Person)*, are represented in CIDOC-CRM version, as *(E55 Type)*. For example, *(E21 Person) has type (P2)* "Viúva" *(E55 Type Marital Status)*.

The person's properties imported from FOAF (above identified) are emphasized in Fig. 2 using dotted line. Similarly, DBpedia properties used are enhanced as dashed line.

This CIDOC-CRM ontology enriched with FOAF and DBpedia elements can describe appropriately the knowledge repository of the Museum of the Person.

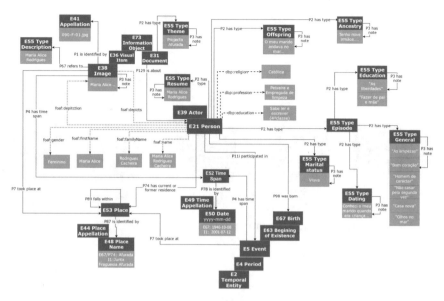

Fig. 2 An instance of CIDOC-CRM/FOAF/DBpedia representation of OntoMP for Maria Cacheira life story (fragment)

3 Proposed Architectures

This section presents a general approach to create a system that builds automatically the Museum of the Person from its repository.

This proposal is defined at an abstract level so that the main architectural blocks and their interactions can be clearly understood; the data flow and the main transformations will be emphasized without technological commitments. We have devised and sketched two possible technical alternatives to implement general architecture. However after describing the general approach, only alternative 1 will be detailed because is the one we chose to refine that architecture.

The general approach, illustrated in Fig. 3, to build the MP comprises: the repository; the Ingestion Function [M1] responsible for getting and processing the input data; a Data Storage (DS) that is the data digital archive; an Ontology to map and link the concepts with the objects stored in (DS); the Generator (M2) to extract data from (DS) and manage the information that will be displayed in Virtual Learning Spaces (VLS) (the final objective of this project) [8].

As said above this general approach has two possible refinements, which are dependent on the (DS). In approach 1 (Figs. 4 and 5) the [DS] is a *TripleStore*, while in approach 2 the (DS) is a *Relational Database*. According to the kind of storage chosen, the ingestion function and the learning spaces generator will require different designs [8].

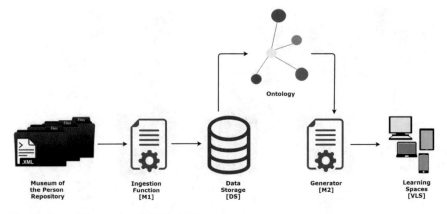

Fig. 3 General Approach to build the MP

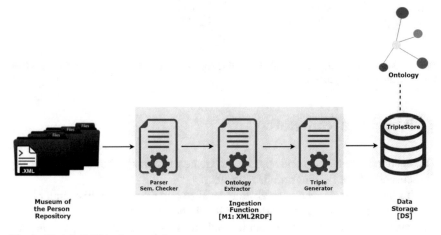

Fig. 4 Module [M1] in Approach 1

The project here imported, we are following the approach 1, so we will describe it in more detail in Sect. 3.1.

In Approach 2 the input XML documents must be converted into SQL to populate the respective database. So, Ingestion Function (M1) is composed of the following components: *Parser and Semantic Checker* that reads the repository documents and extracts the relevant data (annotated in XML), checking their semantic consistency; and *SQL Generator* that generates automatically the SQL statements that insert the retrieved data into the database tables. After the two phases of Ingestion Function [M1] the documents data populate the Relational Database schema, due to the SQL statements generated. As this schema is not directly related to the ontology, in this second approach an explicit mapping is necessary. Making this mapping available, it is possible to resort to CaVa (Criação de Ambientes Virtuais de Aprendizagem) system [4] to build automatically the Virtual Learning Spaces (VLS). Notice that

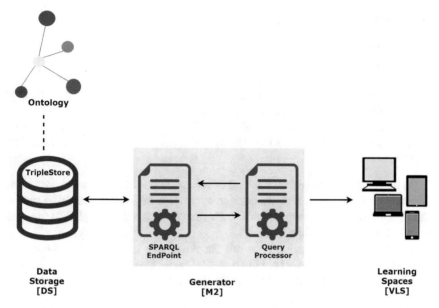

Fig. 5 Module (M2) in Approach 1

only the generator module of CaVa, CaVa^{Gen} will be used in this context. In this case, Generator [M2] is composed of: *DB2Onto Mapping* that associates concepts and relations belonging to the ontology with their respective instances stored in database (it allows to access database tables and fields to get the instances of the ontology concepts); and *CaVa^{Gen}*, that generates automatically the Virtual Learning Spaces from their formal specification based on the ontology.

In this second approach all the work concerned with the query generation according to the exhibition requirements and the answer processing to fulfill the rooms templates is left to CaVa^{Gen}. The only thing that is needed is the specification of the desired learning spaces in CaVa^{DSL}. For more details about this approach please read [8]. The next section will detail approach 1.

3.1 Approach 1

As said above, Approach 1 is based on the decision of using a TripleStore as data storage (DS). According to this decision, Ingestion Module and the Generator Module must be adapted; the first will transform the input XML (eXtensible Markup Language) documents into RDF (Resource Description Framework) triples, and the second will retrieve information from the RDF triples to create the museum web pages.

Figure 4 details the first module that is composed of three blocks [8]:

- *Parser and Semantic Checker* that reads the repository documents and extracts the relevant data (annotated in XML), checking their semantic consistency;
- *Ontology Extractor* that identifies in the extracted data the concepts and relations that belong to the ontology creating in this way an instance of the abstract ontology (in another words, this component populates the ontology);
- *Triple Generator* that converts automatically the ontology triples (created in the preceding block) into triples in RDF notation appropriated to be stored in the (DS) chosen.

At an early stage, to realize the kind of information that contains the documents and how we would represent, we decided to conduct the analysis and extraction of information from documents manually. This means that we accomplished the three phases of Ingestion Module manually.

Among the many existing notations for describing ontologies we chose RDF because we use CIDOC-CRM, FOAF and DBpedia that are described in its original form in RDF. An excerpt of the RDF triples built by hand is shown in Listing 1.

Listing 1 Fragment of the RDF Triples for Maria Cacheira life story

```
1   <!-- Description Interviewed 1 -->
2   <rdf:Description rdf:about="&ecrm;Interviewed_1">
3       <rdf:type
        rdf:resource="&ecrm;E21_Person"/>
4       <rdf:type rdf:resource=
        "http://dbpedia.org/ontology/Person"/>
5       <rdf:type rdf:resource="&foaf;Person"/>
6
7       <foaf:firstName
        rdf:datatype="&xsd;string">Maria
        Alice</foaf:firstName>
8       <foaf:name
        rdf:datatype="&xsd;string">Maria Alice
        Rodrigues Cacheira</foaf:name>
9       <foaf:familyName
        rdf:datatype="&xsd;string"> Rodrigues
        Cacheira</foaf:familyName>
10      <P98i_was_born rdf:resource="&ecrm;B1"/>
11
12      <foaf:gender rdf:datatype="&xsd;string">
        Feminino</foaf:gender>
13      <foaf:depiction rdf:resource=
        "&ecrm;I1_Interviewed_1"/>
14
15      <dbp:profession
        rdf:datatype="&xsd;string"> Peixeira e
        empregada de limpeza</dbp:profession>
16      <dbp:religion rdf:datatype="&xsd;string">
        Catolica</dbp:religion>
17      <dbp:education
        rdf:datatype="&xsd;string"> Sabe ler e escrever
```

```
             (quarta classe)</dbp:education>
           </rdf:Description>
18
19   <!-- Event Birth Interviewed 1   (B1) -->
20   <rdf:Description rdf:about="&ecrm;B1">
21          <rdf:type rdf:resource="&ecrm;E67_Birth"/>
22          <P98_brought_into_life rdf:resource=
           "&ecrm;Interviewed_1"/>
23          <P4_has_time-span
           rdf:resource="&ecrm;TS1"/>
24          <P7_took_place_at
           rdf:resource="&ecrm;PL1"/>
25   </rdf:Description>
26
27   <!-- Description Photo Interviewed 1 (I1) -->
28   <rdf:Description
           rdf:about="&ecrm;I1_Interviewed_1">
29          <rdf:type rdf:resource="&ecrm;E38_Image"/>
30          <rdf:type rdf:resource="&foaf;Image"/>
31          <foaf:depicts
           rdf:resource="&ecrm;Interviewed_1"/>
32          <P67_refers_to rdf:resource=
           "&ecrm;I1_Description_Interviewed_1"/>
33          <P3_has_note rdf:datatype="&xsd;string">
           Maria Alice Rodrigues Cacheira</P3_has_note>
34          <P1_is_identified_by rdf:resource=
           "&ecrm;090-F-01.jpg"/>
35          <P4_has_time-span
           rdf:resource="&ecrm;TS7"/>
36          <P7_took_place_at
           rdf:resource="&ecrm;PL8"/>
37   </rdf:Description>
```

The triple fragment shown in Listing 1 contains information about life story of Maria Cacheira. The biographic information about Maria Cacheira, as name (first, last name and full name), birth, sex, photo, profession, religion, and education is displayed in first section (line 1–17). The birth event of Maria Cacheira, date and place of it, is described in the second section (line 19–25). Finally, the last section (line 27–37) contains specific information about the photo of the interviewed, such as description, legend, file, date and place.

The next step was to use the W3C online tool RDF Validator[2] to validate the handwritten triples to ensure that the very long textual description produced contains no errors. RDF Validator checks the consistency of the triple RDF and displays them in a table with three columns 'subject, predicate and object'. After loading our RDF file we got, as feedback, the information *"VALIDATION RESULTS: Your RDF document validated successfully"* that is just what we want to get from that tool.

The next step, after the successful validation, was to store the triples in a data set, a RDF database, called Apache Jena TDB.

---

[2]https://www.w3.org/RDF/Validator/.

TDB is a component of Jena (free and open source Java framework for building Semantic Web and Linked Data applications) for RDF storage and query, and can be used as a high performance RDF store on a single machine. A TDB store can be accessed and managed with the provided command line scripts and via the Jena API. Apache Jena Fuseki component provides a SPARQL server to be used with TDB [14].

By performing these three phases of the Ingestion Function (M1), we understand how to make the extraction and analysis of semantic concepts and how to convert the triple ontology in RDF triples. You also realize that it is a very time consuming work to be done manually for all documents in the repository study, then we decide to create a tool to do these three phases automatically. This tool will be described in detail in Sect. 4.

As the mapping between the domain ontology (previously defined) and the data extracted from the repository is automatically built by construction in the second block, above, there is no need to create explicitly this mapping. It means that the Generator [M2] can access directly the storage to obtain the conceptual information necessary to create the exhibition rooms [8].

To display in the Virtual Learning Spaces (VLS) the information stored in (DS)–TripleStore, the (VLS) Generator needs to send queries and process the returned data.

Figure 5 shows the second module [M2] (the Generator) that is composed of two blocks [8]:

- *SPARQL Endpoint* that receives and interprets the SPARQL queries, accesses the TripleStore and returns the answers. For this, it is necessary to resorted to a SPARQL Endpoint. The SPARQL endpoint used was Apache Jena Fuseki (version 2.0).

  Apache Jena Fuseki is a SPARQL server, that can run as an operating system service, as a Java web application (WAR file), and as a standalone server. Fuseki is tightly integrated with TDB to provide a robust, transactional persistent storage layer, and incorporates Jena text query and Jena spatial query [15]. To check if we could extract information from the created ontology, we built some queries. An example of a query that has been built is the one listed below to find the name of the Interviewed of a given sex and residence.

Listing 2 Query SPARQL: Interviewed by sex and residence

```
1  PREFIX : <http://erlangen-crm.org/150929/>
2  PREFIX foaf: <http://xmlns.com/foaf/0.1/>
3  PREFIX dbp: <http://dbpedia.org/ontology/>
4  PREFIX rdf:
      <http://www.w3.org/1999/02/22-rdf-syntax-ns#>
5  PREFIX xsd: <http://www.w3.org/2001/XMLSchema#>
6
7  SELECT DISTINCT ?name
8
9  WHERE {
10
```

```
11   ?pessoa a :E21_Person;
12      :P129_is_subject_of ?doc;
13      foaf:name ?name;
14      foaf:gender"sexInterviewed"^^xsd:string ;
15      :P74_has_current_or_former_residence ?place.
16        ?place :P87_is_identified_by ?parish .
17        ?parish :P3_has_note
        "Name-Residence"^^xsd:string .
18
19   } ORDER BY ?name
```

The code block between lines 11 and 17 of Listing 2 is designed to search for all respondents *(E21_Person)* of a given sex *(foaf:gender)* who live in a given location *(:P74_has_current_or_former_residence)*. For example it can be instantiated to, list all the female respondents living in Afurada. The property *foaf:name* describes the full name of each respondent.

For more information on the results and executed queries, please see: http://npmp.epl.di.uminho.pt.

- *Query Processor* generates the SPARQL queries according to the exhibition room requirements, sends them to the SPARQL Endpoint and after receiving the answer, combines the returned data to set up the Virtual Learning Spaces (VLS).

We created a Python script that generates SPARQL queries according to the requirements of each exhibition room, sends them to the Fuseki (SPARQL Endpoint) and after receiving the answer, combines the data returned to configure the Virtual Learning Spaces (VLS). This Python script also includes HTML (Hyper Text Markup Language) and CSS (Cascading Style Sheets) to create and format the web page.

Finally, to exhibit the life stories that are the objects of the Museum of the Person, the web pages were built (Virtual Learning Spaces).

Figure 6 displays the page where the museum visitor can perform the SPARQL queries.

The answer to the query referred in Listing 2 (Interviewed by sex and residence) is shown in the Fig. 7.

In this approach, each Virtual Learning Space (a museum's exhibition room) is built fulfilling a web page template with the concrete data retrieved from the data store.

4 XML Repository and Ontology Extraction

As mentioned in Sect. 3.1, we initially performed the extraction and analysis of the semantic concepts, and convert manually those triples into RDF triples. After understanding the structure of the documents, the information they contain and how to convert the ontology triples into RDF triples, we decided to develop a text filter able

Fig. 6 Consulting the Museum of the person repository

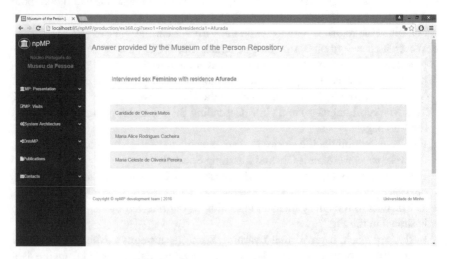

Fig. 7 Response to the SPARQL query: *Interviewed by sex and residence*

to scan all files that compose an interview (BI, Edited Interview and Photograph Captions), extract relevant information and convert into a single RDF triples file.

To develop the text filter we used the compilers generator system AnTLR (Another Tool for Language Recognition) [16] integrated in AnTLRWorks tool, version 2.1, a plugin for NetBeans IDE. AnTLR is a powerful parser generator for reading, and processing structured text files; so it is extensively used to build language-based tools, and frameworks.[3]

---

[3] http://www.antlr.org/index.html.

In our case, from a set of Regular Expressions (RE), AnTLR will generate a Lexical Analyzer that realizes the desired text filter for data extraction.

That text filter, or extractor, will accept an input document, like the one exemplified in Listing 3, and after analyzing and processing it will output a RDF description, like the one shown in Listing 4.

Listing 3 An XML input document

```
1   <?xml version="1.0" encoding="ISO-8859-1"?>
2   <fotos>
3       <foto ficheiro="090-F-01.jpg">
4           <quem>Maria Alice Rodrigues
        Cacheira</quem>
5           <onde>Junta de Freguesia da Afurada</onde>
6           <quando data="2001-07-12"/>
7       </foto>
8       <foto ficheiro="090-F-02.jpg">
9           <quem>Maria Alice Rodrigues
        Cacheira</quem>
10          <onde>Junta de Freguesia da
        Afurada</onde><
11          <quando data="2001-07-12"/>
12      </foto>
13  </fotos>
```

Listing 4 An RDF output document

```
1    <rdf:Description rdf:about="&ecrm;090-F-01.jpg">
2        <rdf:type rdf:resource="&ecrm;E41_Appellation"/>
3    </rdf:Description>
4
5    <rdf:Description rdf:about="&ecrm;I0_Interviewed_1"/>
6        <rdf:type rdf:resource="&ecrm;E38_Image"/>
7        <rdf:type rdf:resource="&foaf;Image"/>
8        <foaf:depicts rdf:resource="&ecrm;Interviewed_1"/>
9        <P67_refers_to
         rdf:resource="&ecrm;I0_Description_Interview_1"/>
10       <P1_is_identified_by rdf:resource="&ecrm;090-F-01.jpg"/>
11       <P4_has_time-span rdf:resource="&ecrm;TS1"/>
12       <P7_took_place_at rdf:resource="&ecrm;PL1"/>
13   </rdf:Description>
14
15   <rdf:Description rdf:about="&ecrm;2001-07-12">
16       <rdf:type rdf:resource="&ecrm;E49_Time_Appellation"/>
17   </rdf:Description>
18
19   <rdf:Description rdf:about="&ecrm;TS1">
20       <rdf:type rdf:resource="&ecrm;E52_Time-Span"/>
21       <P78_is_identified_by rdf:resource="&ecrm;2001-07-12"/>
22   </rdf:Description>
23
24   <rdf:Description rdf:about="&ecrm;PL1">
25       <rdf:type rdf:resource="&ecrm;E53_Place"/>
26       <P87_is_identified_by rdf:resource="&ecrm;Place1"/>
27   </rdf:Description>
28
29   <rdf:Description rdf:about="&ecrm;Place1">
30       <rdf:type rdf:resource="&ecrm;E48_Place_Name"/>
```

```
31        <P3_has_note rdf:datatype="&xsd;string">Junta de Freguesia da
             Afurada</P3_has_note>
32    </rdf:Description>
33
34    <rdf:Description rdf:about="&ecrm;I0_Description_Interview_1">
35        <rdf:type rdf:resource="&ecrm;E55_Type"/>
36        <P2_has_type rdf:resource="&ecrm;Description"/>
37        <P3_has_note rdf:datatype="&xsd;string">Maria Alice Rodrigues
             Cacheira</P3_has_note>
38    </rdf:Description>
39
40    <rdf:Description rdf:about="&ecrm;090-F-02.jpg">
41        <rdf:type rdf:resource="&ecrm;E41_Appellation"/>
42    </rdf:Description>
43
44    <rdf:Description rdf:about="&ecrm;I1_Interviewed_1"/>
45        <rdf:type rdf:resource="&ecrm;E38_Image"/>
46        <rdf:type rdf:resource="&foaf;Image"/>
47        <foaf:depicts rdf:resource="&ecrm;Interviewed_1"/>
48        <P67_refers_to
             rdf:resource="&ecrm;I1_Description_Interview_1"/>
49        <P1_is_identified_by rdf:resource="&ecrm;090-F-02.jpg"/>
50        <P4_has_time-span rdf:resource="&ecrm;TS1"/>
51        <P7_took_place_at rdf:resource="&ecrm;PL1"/>
52    </rdf:Description>
```

That automatic transformation is obtained using a specification (an *AnTLR Lexer grammar*) illustrated in Listing 5. The fragment shown in the referred listing is a sequence of transformation rules that corresponds to the beginning of the global specification (the specification part not included will be discussed below). Each rule has a name and a pair composed of a Regular Expression (RE) and a Semantic Action (SE) written in Java. A rule is interpreted from left to right: if the Regular Expression is found in the input, then the corresponding Semantic Action is triggered. The RE defines the text pattern that shall be find in the input, and the SE specifies how the concrete text found shall be transformed.

Moreover, AnTLR lets the programmer to set up modes that group the specific rules to address each sub block in the input file.

In Listing 5 it can be seen the three rules (namely, *Cabec, Fotos e MP*) corresponding to the three input files (*BI, Photography Captions,* and *Edited Interview*), respectively. When the extractor reads a XML tag definning the beginning of one of these three documents, it enters a special AnTLR mode to process that document's content.

Listing 6 shows the main mode to process the *Photography Caption* XML documents. The listing illustrates the general approach adopted: when a block opening tag is found, the appropriate mode is entered to consume the block contents; when the block closing tag is found, the processor exits the mode and returns to the initial mode.

The four auxilairy modes, called from the main one (see lines 13–19), contains the specific rules used to extract information from the four main blocks of the *Photography Captions* input document. Listing 7 contains the rules (just a fragment is shown) executed at the end of the processing (mode activated at line 21) to print out the RDF triples built in the internal representation. This grammar fragment is actually responsible for the generation of the RDF output file.

Listing 5 XML2RDF Lexer Grammar for AnTLR

```
1   lexer grammar XML2RDF;
2
3   Cabec   :   '<'[Bb][Ii]'>'                     ->
        mode(sBI)
4           ;
5   Fotos   :   '<'[Ff][Oo][Tt][Oo][Ss]'>'    -> mode
        (sFOTOS)
6           ;
7   MP      :   '<'[Mm][Pp]'>'                     -> mode
        (sMP)
8           ;
9   Default:    .              { ; }
10          ;
11
12          ...
13  ........Modes specification........
14          ...
```

Listing 6 Lexer Grammar Photos main Mode

```
1   mode sFOTOS;
2   GetSFOTOS           :   '<foto'      -> mode(sFOTO)
3                          ;
4   OutFOTOSSAVE        :   '</fotos>'  ->
        mode(DEFAULT_MODE)
5                          ;
6   DefaultsFOTOS   :   .        { ; }
7                          ;
8
9
10  mode sFOTO;
11  GetFOTO     :   [ ]+'ficheiro="'     -> mode
        (sFICHEIRO)
12                          ;
13  GetQUEM     :   '<quem>'              -> mode (sQUEM)
14                          ;
15  GetQUANDO   :   '<quando'             -> mode
        (sQUANDO)
16                          ;
17  GetFACTO    :   '<facto>'             -> mode
        (sFACTO)
18                          ;
19  GetONDE     :   '<onde>'              -> mode (sONDE)
20                          ;
21  OutFOTOS    :   '</'                  -> mode
        (sPRINTTUDO)
22                          ;
23
24  DefaultsFOTO :   .      { ; }
25                          ;
```

Listing 7 Lexer Grammar Print Mode

```
1   mode sPRINTTUDO;
2   GetsPRINTTUDO    :'foto'      {
3
        pessoa.AddImage("I"+newCountKeyFicheiro+"_Interviewed_"+
        countinterview);
4
5                                       System.out.print("<rdf:Description
        rdf:about=\"&ecrm;");
6
        System.out.println(ficheiro+"\">");
7                                       System.out.println("\t<rdf:type
        rdf:resource=\"&ecrm;E41_Appellation\"/>");
8
        System.out.println("</rdf:Description>\n\n");
9
        System.out.println("<rdf:Description
        rdf:about=\"&ecrm;I"+newCountKeyFicheiro+"_Interviewed_"+
        countinterview+"\"/>");
10                                      System.out.println("\t<rdf:type
        rdf:resource=\"&ecrm;E38_Image\"/>");
11                                      System.out.println("\t<rdf:type
        rdf:resource=\"&foaf;Image\"/>");
12
        System.out.println("\t<foaf:depicts
        rdf:resource=\"&ecrm;Interviewed_"+countinterview+"/>");
13
14      if(!quem.equals("")){
15              System.out.println("\t<P67_refers_to
        rdf:resource=\"&ecrm;I"+newCountKeyFicheiro+"_Description_I
16
17  nterview_"+ countinterview+"\"/>");}
18
19      if(!facto.equals("")){
20              System.out.println("\t<P3_has_note
        rdf:datatype=\"&xsd;string\">"+facto+"</P3_has_note>"); }
21
        System.out.println("\t<P1_is_identified_by
        rdf:resource=\"&ecrm;"+ficheiro+"\"/>");
22
23
24      ...
25
26
27  OutsPRINTTUDO : '>'       -> mode(sFOTOS)
28          ;
```

5   Conclusion

This paper describes the creation of a virtual museum to exhibit people's life stories, called the Museum of the Person (MP). Museum of the Person[4] was born in Brazil, São Paulo, in 1991, created by a group of historians who decided to build the country's history using testimonials of ordinary people [17]. Our work concerns the

---

[4]Accessible at: http://www.museudapessoa.net.

Portuguese branch of such network of life stories museums, npMP. From the life stories of individuals, the objective is to write up the stories of families, communities, or institutions.

After analyzing the documents that make up the repository, we designed OntoMP, an ontology for the Museum of the Person. The next stage after the validation and tuning of OntoMP was to describe it in a standard ontology format used for museums, CIDOC-CRM (CIDOC Conceptual Reference Model) complemented with some pertinent FOAF and DBpedia concepts and properties.

In this paper we propose a general architecture to build a software platform to create the museum's virtual exhibition rooms, as web pages, extracting information from the museum's repository. To implement the overall architecture outlined there are two possible alternative techniques. However, to refine this architecture, we chose approach 1. One approach uses a TripleStore to archive the ontology instances and resorts to SPARQL technology to query the repository and obtain the information that will be exhibited. The other approach uses a Relational Database as archive and reuses CaVa framework to extract and display the information. CaVa is a novel proposal under development in the context of the PhD project of one of the authors, and our first objective was to use npMP as a second case study to test that framework.

After implementing the approach 1, we came to the conclusion that to implement the first module (Ingestion Function) manually is a very lengthy process. So we decided to create a text filter to perform the three phases of this module automatically, as was discussed in the article.

As future work we intend to refine the filter in some aspects, particularly in the recursive episodes, among others in order to be possible to deal with all the documents stored in our present repository.

Acknowledgements This work has been supported by COMPETE: POCI-01-0145-FEDER-007043 and FCT – Fundação para a Ciência e Tecnologia within the Project Scope:UID/CEC/00319/2013. The work of Ricardo Martini is supported by CNPq, grant 201772/2014-0.

References

1. Almeida, J.J., Rocha, J.G., Henriques, P.R., Moreira, S., Simões, A.: Museu da Pessoa–arquitectura. In: Encontro Nacional da Associação de Bibliotecários, Arquivista e Documentalistas, ABAD'01. BAD (2001)
2. Rodrigues, B.C., Crippa, G.: Novas Propostas e Desafios Das Mediações Culturais em Museus Virtuais. In: El Pensamiento Museuloógico Contenporá neo. O Pensamento Museulógico Contemporâneo, pp. 599–608. ICOM (2011)
3. Philip, B.: Stafford. Museum of person, Technical report (2015)
4. Araújo, C.: An Ontology for the Museum of the Person Combining CIDOC-CRM with FOAF. Universidade do Minho, Msc pre-thesis (2016)
5. Martini, R.: Formal Description and Automatic Generation of Learning Spaces based on Ontologies. Universidade do Minho, Ph.D. pre-thesis (2015)
6. Simões, A., Almeida, J.J.: Histórias de Vida + Processamento Estrutural = Museu da Pessoa. In: XATA 2003 — XML: Aplicações e Tecnologias Associadas, pp. 16. Braga, Portugal (2003). UM

7. Martini, R.G., Araújo, C., Almeida, J.J., Henriques, P.R.: New advances in information systems and technologies: volume 2. In: chapter OntoMP, An Ontology to Build the Museum of the Person, pp. 653–661. Springer International Publishing, Cham (2016)
8. Araújo, C., Martini, R.G., Henriques, P.R., Almeida, J.J.: Architectural approaches to build the museum of the person. In: Rocha, Á., Reis, L.P., Cota, M.P., Suárez, O.S., Gonçalves, R. (eds.) Sistemas y Tecnologías de Información—Atas da 11ª Conferência Ibérica de Sistemas e Tecnologias de Informação, volume Vol. I — Artículos de la Conferencia, pp. 383–388. AISTI–Associação Ibérica de Sistemas e Tecnologias de Informação, June 2016
9. Martini, R.G., Araújo, C., Librelotto, G.R., Henriques, P.R.: New advances in information systems and technologies. In: chapter A Reduced CRM-Compatible Form Ontology for the Virtual Emigration Museum, pp. 401–410. Springer International Publishing, Cham (2016)
10. ICOM/CIDOC. Definition of the CIDOC Conceptual Reference Model. Technical report, ICOM/CIDOC, May 2015
11. Allemang, D., Hendler, J.: Semantic Web for the Working Ontologist: Effective Modeling in RDFS and OWL. Elsevier Science (2011)
12. Al-Mukhtar, M.M.A., Al-Assafy, A.T.A.: The implementation of foaf ontology for an academic social network. Int. J. Sci. Eng. Comput. Technol. 4(1), 10 (2014)
13. Dbpedia. Ontology. http://wiki.dbpedia.org/ (2016). Accessed 15 June 2016
14. APACHE JENA. TDB. https://jena.apache.org/documentation/tdb/index.html (2016). Accessed 01 June 2016
15. APACHE JENA. Apache Jena Fuseki. https://jena.apache.org/documentation/fuseki2/index.html (2016). Accessed 01 June 2016
16. ANTLR. ANTLR. http://www.antlr.org/ (2016). Accessed 14 Sept 2016
17. Worcman, K.: The museum of the person. In: Virtual Museums, vol. 57, no. 3. ICOM (2004)

Web-Based Decision System
for Distributed Process Planning
in a Networked Manufacturing
Environment

V.K. Manupati, P.K.C. Kanigalpula, M.L.R. Varela,
Goran D. Putnik, A.F. Araújo and G.G. Vieira

Abstract Distributed manufacturing sector is increasingly enabling the Web-enabled services due to the advancements in Information technology and pervasive applications of recently advanced manufacturing systems. In this paper, world wide web (WWW) collaborative model is developed, and architecture with the web enabled service system for effective integration of distributed process planning and scheduling is proposed to assists the geographically distributed enterprises located in the context of networked manufacturing for effective coordination and collaboration. To validate the feasibility of the proposed approach, a case study has been presented and found that the proposed method and developed tool offers some benefits such as high interoperability, openness, cost-efficiency, and production scalability.

V.K. Manupati
Division of Manufacturing, School of Mechanical Engineering,
VIT University, Vellore, Tamil Nadu, India
e-mail: manupativijay@gmail.com

P.K.C. Kanigalpula
Department of Mechanical Engineering, IIT Kharagpur,
Kharagpur, West Bengal, India
e-mail: kpkchakravarthy@gmail.com

M.L.R. Varela (✉) · G.D. Putnik · A.F. Araújo · G.G. Vieira
Department of Production and Systems,
School of Engineering, University of Minho, Guimarães, Portugal
e-mail: leonilde@dps.uminho.pt

G.D. Putnik
e-mail: putnikgd@dps.uminho.pt

A.F. Araújo
e-mail: dricafaraujo@hotmail.com

G.G. Vieira
e-mail: gaspar_vieira@hotmail.com

© Springer International Publishing AG 2018
Á. Rocha and L.P. Reis (eds.), *Developments and Advances in Intelligent
Systems and Applications*, Studies in Computational Intelligence 718,
DOI 10.1007/978-3-319-58965-7_8

1 Introduction

Recent advancements in information and communication technologies enhanced the latest manufacturing systems capability by increasing the accessibility of Internet technology. Digitalization and networking of enterprises bring the challenges of accessibility and interoperability in a distributed environment. Therefore, advance and pervasive applications are required to enhance the interoperability and also to support the distributed collaboration. However, manufacturing systems are advancing towards supporting distributed and collaborative activities in a distributed manufacturing environment. Due to light weight, ease of accessibility, and high flexibility the World Wide Web (web or WWW) technology offers tremendous opportunities for the sharing of information among enterprises that are globally distributed.

A web application is defined as any software application that depends on the web for its correct execution [1]. In this paper, the integration of process planning and scheduling in connection with web-enabled service is made available on the Internet platform to support the networked manufacturing services. Wagner et al. [2] proposed a remote analysis of CAD models for the exchange of geometric data. In their work, they tried its implementation over industry environment and then proved the model effectiveness. Tele-manufacturing a service-based paradigm developed by Bailey [3, 4] to provide rapid prototyping services on the internet. Sung et al. [5] proposed a web-based decision system for the integration of product design and process planning using CyberCut experiment based on Java-based programming. Mervyn et al. [6] developed a Web-based fixture design system where information and knowledge among modules are exchange with proposed XML format.

Several contributions i.e. CyberCut and MADEFAST experiments in the area of production scheduling problem were developed, but they were not interoperable. Several note worthy efforts have been observed in distributed object paradigms such as Microsoft Distributed Component Object Model (DCOM), OMG's Common Object Request Broker Architecture (CORBA), JavaSoft's Java/Remote Method Invocation (Java/RMI) [7]. It is found that many effective knowledge integration methods for product design, process, and manufacturing have been developed [8–12]. Out of all these integration methods, Knowledge Interchange Format (KIF) for KQML (Knowledge Query and Manipulation Language) and STEP (Standard for the Exchange of Product model data). Although many standard formats available, still there is a gap on the volume of data that need to be exchanged between different distributed systems [13–15] to develop. Varela et al. [16] developed an effective web-based decision support system for effective integration of concurrent manufacturing activities.

The above-mentioned issues have not been dealt adequately thus in this research work; we have developed a web-based decision support system (WBDSS) which can be acting as an interactive medium for concurrent activities among various manufacturing activities. Subsequently, the proposed architecture shows how the interactions between different modules can be processed. Moreover, with a flow chart, the step by step procedure of different functionalities is detailed. Finally, with a case study the proposed approach for the design and development of an inter-active distributed manufacturing environment for internet users has been organized and achieved the concurrent engineering activities effectively.

This paper has been classified into six major section. Section 2 described network manufacturing system its characteristics with an architecture and detailed its working procedure. In Sect. 3 the five functional modules of developed world wide web modeling schema are presented. Section 4 presented the framework of the proposed web enabled service system, and its detailed procedure is discussed. The paper ends with Sect. 5 with a conclusion and proposes future works related to the study.

2 Network Manufacturing System Description

A networked manufacturing system (NMS) can be defined as a manufacturing-oriented network that has the capability to encapsulate the manufacturing enter-prises' information to provide the manufacturing services. In Fig. 1 the architecture of NMS that starts with a request from the customer user whose product task can be handled by the web-based manufacturing service. In general, any manufacturing service has two different modes to serve viz. Customer User (CU) and Enterprise

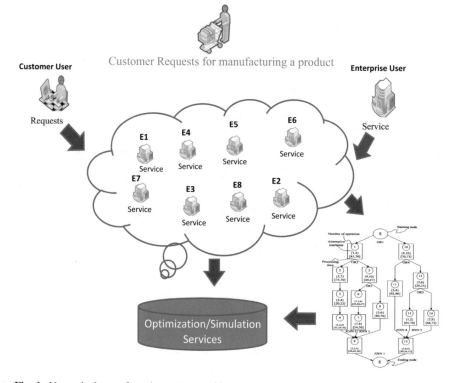

Fig. 1 Networked manufacturing system architecture

user (EU). CU is defined as a customer/organization who can accept the requests of the customers to process it depend on the product category. However, these accepted requests are further transferred to their data base for processing it with respective facilities. Where the support of web-based decision system plays an important role to provide a feasible solution in an effective manner.

On the other side, the functionality of EU has the capability to service the requested product by itself. In real life cases, it is not possible to perform all the tasks by the single enterprise; this is same for even EU. The EU can serve as a centralize organization where the initiation to interact with the customer, and collaborating with other related enterprises as a coordinator. In this paper, we have considered the User mode to carry the research. Here, after finding the necessary product data and the enterprises' information, an effective approach to describe the manufacturing functions requirements and their implementation on networked manufacturing environment is accomplished.

3   Modeling Schema for the Functions of Web-Based System

In this section, the four functional modules is deployed and their detailed structure with functionalities are shown in Fig. 2. The Module 1 represents the modeling phase where sketches of 2D/3D models for the desired product is created. In this

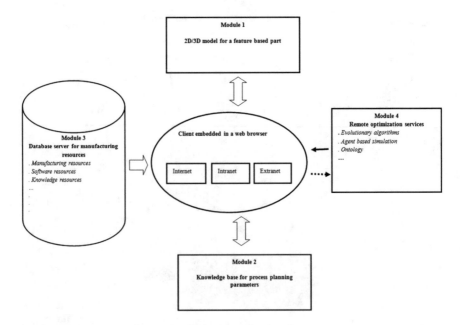

Fig. 2 Modeling schema for the functions of web-based system

phase, the front-end-client embedded in a web browser provides users with various functionalities such as visualization of design models, invoking remote process optimization services, and an interface to data base and knowledge servers to store and retrieve information needs.

In Module 2, the detailed information of a product/part and its related information for processing such as operations, machines, processing times, process plans, etc. can be considered. In other words, the knowledge base of the above requirements is useful to determine the decision makers criteria. Module 3 indicates a database server where manufacturing resources information, i.e., available machines, tools, TAD and their costs utilized for determining process plans of a design model is presented. In the final Module 4 several remote optimization/ simulation services for optimal process plans as analysis Servlet that would respond through a web server in the online invocation and evaluation.

4 Flowchart of the Proposed Method with the Web Enabled Service System Architecture

In this section, the architecture of the proposed web enabled service system is detailed, and the execution of the above-mentioned four modules in a collaborative way is shown in Fig. 3. Primarily, through survey the required information is gathered, and it is employed for the database development. Here, we have used manufacturing industry as a case where feature based products' data are major

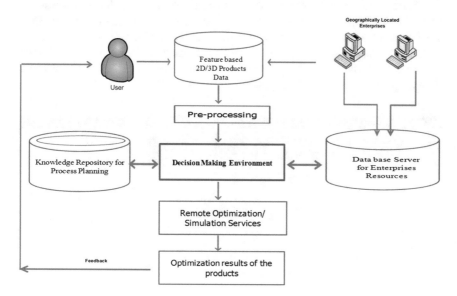

Fig. 3 Flow chart of the web-enabled system architecture

resource data. After that, the information is converted into Ontology files to generate knowledge from the information with the help of decision manager and then stored it in the knowledge repository. These generated files with knowledge can be easily converted into XML files where the flexibility to transfer and exchange the information between distributed manufacturing resources. The flowchart of various modules and their integration with the developed DSS is depicted in Fig. 3.

After data pre-processing, decision-making environment initiate selection of appropriate algorithms, approaches or methods for solving the concerned problem. Execution of the modules on problem nature takes place according to the selection of the user interface. The developed static web service system environment acts as an interface mechanism with the remote optimization/simulation services for the execution of the user requirements. The obtained results from remote optimization services are presented in the form of Gantt charts, bar charts, figures, tables, etc. and then transferred to the user. Another important functionality of this DSS is, it can provide a feedback message to the user in real time, if further improvement or manipulation of the results is required.

5 Web-Based Decision Supports System Tool

In this section, with developed tool, the execution of the proposed manufacturing system is implemented. A snapshot of the web based decision system tool is presented in Fig. 4 from which design and development of an interactive distributed manufacturing environment for the internet users can be successfully achieved.

With this tool the execution of the four developed modules is successfully executed. As the tool is more interactive if any data is required and it should be filled certain internal modules presented in the tool can take care for the execution. Several insights reveal that from the developed methodologies and tool for networked manufacturing environment offers some benefits such as high interoperability, openness, cost-efficiency, and production scalability.

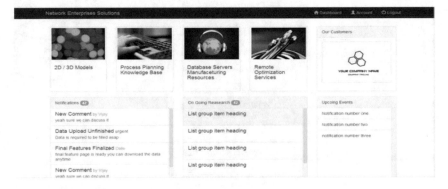

Fig. 4 Snapshot of the developed web-based decision system tool

6 Conclusion and Future Work

The networked manufacturing system is considered as an advanced manufacturing paradigm where the enterprises can run their businesses with recently advanced network and communication technologies. It has several capabilities to serve as a potential system for integrated functions with design, manufacturing, and management. In this research work, with a case study, an attempt has been made to develop a different approach to a flowchart, architecture, and a web-based decision support system to support networked manufacturing system for process planning and scheduling integration. The developed platform behaves as a convenient platform for the users to view, model, evaluate for invoking remote process planning optimization services. The proposed approach is a generalized one. Therefore it is possible to generate flexible process plans for any type of complex parts once the process capability information is updated in the system. However, the present system is limited to prismatic parts in manufacturing. The developed intelligent and interactive distributed process planning system is user-friendly, flexible and expandable in nature. In near future, it can be extended to all kinds of manufacturing parts and distributed problems in supply chain management.

References

1. Gellersen, H.W., Gaedke, M.: Object-oriented web application development. IEEE Internet Comput. **3**(1), 60 (1999)
2. Wagner, R., Castanotto, G., Goldberg, K.: FixtureNet: interactive computer-aided design via the World Wide Web. Int. J. Hum Comput Stud. **46**(6), 773–788 (1997)
3. Wright, D.T., Burns, N.D.: Cellular Green-Teams in global network organisations. Int. J. Prod. Econ. **52**(3), 291–303 (1997)
4. Bailey, M.W., VerDuin, W.H.: FIPER: An Intelligent System for the Optimal Design of Highly Engineered Products, pp. 467–477. NIST Special Publication SP (2001)
5. Sung, Y.T., Chang, K.E., Chiou, S.K., Hou, H.T.: The design and application of a web-based self-and peer-assessment system. Comput. Educ. **45**(2), 187–202 (2005)
6. Mervyn, F., Bok, S.H., Nee, A.Y.C.: Development of an Internet-enabled interactive fixture design system. Comput. Aided Des. **35**(10), 945–957 (2003)
7. Huang, G.Q., Mak, K.L.: Design for manufacture and assembly on the Internet. Comput. Ind. **38**(1), 17–30 (1999)
8. Ozman, M.: Knowledge integration and network formation. Technol. Forecast. Soc. Chang. **73**(9), 1121–1143 (2006)
9. Bless, P.N., Klabjan, D., Chang, S.Y.: Heuristics for automated knowledge source integration and service composition. Comput. Oper. Res. **35**(4), 1292–1314 (2008)
10. Bombardier, V., Mazaud, C., Lhoste, P., Vogrig, R.: Contribution of fuzzy reasoning method to knowledge integration in a defect recognition system. Comput. Ind. **58**(4), 355–366 (2007)
11. Gardner, S.P.: Ontologies and semantic data integration. Drug Discov. Today **10**(14), 1001–1007 (2005)
12. Kwon, O., Kim, K.Y., Lee, K.C.: MM-DSS: integrating multimedia and decision-making knowledge in decision support systems. Expert Syst. Appl. **32**(2), 441–457 (2007)

13. Uschold, M., Grueninger, M.: Ontologies: principles, methods, and applications. Knowl. Eng. Rev. **11**(2), 93–155 (1996)
14. Lin, H.K., Harding, J.A.: A manufacturing system engineering ontology model on the semantic web for inter-enterprise collaboration. Comput. Ind. **58**(5), 428–437 (2007)
15. Daconta, M.C., Obrst, L.J., Smith, K.T.: The Semantic Web: A Guide to the Future of XML, Web Services, and Knowledge Management. Wiley (2003)
16. Varela, M.L.R., Araujo, F., Putnik, G.D., Manupati, V.K., Anirudh, K.V.: Web-based decision system for effective process planning in networked manufacturing system. In: 11th Iberian Conference on Information Systems and Technologies (CISTI), pp. 1–7 (2016). doi:10.1109/CISTI.2016.7521371

Managing the Lifecycle of Security SLA Requirements in Cloud Computing

Marco Antonio Torrez Rojas, Fernando Frota Redígolo, Nelson Mimura Gonzalez, Fernando Vilgino Sbampato, Tereza Cristina Melo de Brito Carvalho, Kazi Walli Ullah, Mats Näslund and Abu Shohel Ahmed

Abstract One of the major barriers for full adoption of cloud computing is the security issue. As the cloud computing paradigm presents a shared management vision, it is important that security requirements are addressed inside the Service Level Agreements (SLAs) established between cloud providers and consumers, along with the tools and mechanisms necessary to deal with these requirements. This work aims at proposing a framework to orchestrate the management of cloud services and security mechanisms based on the security requirements defined by a SLA, in an automated manner, throughout their lifecycles. In addition, the integration of the framework with a cloud computing solution is presented, in order to demonstrate and validate the framework support throughout SLAs lifecycle phases.

M.A.T. Rojas (✉) · F.F. Redígolo · N.M. Gonzalez · F.V. Sbampato ·
T.C.M. de Brito Carvalho
Escola Politécnica, University of São Paulo, São Paulo, Brazil
e-mail: matrojas@larc.usp.br

F.F. Redígolo
e-mail: fernando@larc.usp.br

N.M. Gonzalez
e-mail: nmimura@larc.usp.br

F.V. Sbampato
e-mail: fsbampato@larc.usp.br

K.W. Ullah · A.S. Ahmed
Ericsson Research, Jorvas, Finland
e-mail: kazi.wali.ullah@ericsson.com

A.S. Ahmed
e-mail: ahmed.shohel@ericsson.com

M. Näslund
Ericsson Research, Stockholm, Sweden
e-mail: mats.naslund@ericsson.com

© Springer International Publishing AG 2018
Á. Rocha and L.P. Reis (eds.), *Developments and Advances in Intelligent Systems and Applications*, Studies in Computational Intelligence 718,
DOI 10.1007/978-3-319-58965-7_9

1   Introduction

Cloud computing providers deliver on-demand services based on shared and distributed infrastructures. This delivery approach is characterized by the deployment and service models adopted by providers (i.e., whether a public/private/community cloud and a Sofware-, Platform- or Infrastructure-as-a-Service offering). The shared responsibilities and the segregation of the roles between cloud providers and consumers are also consequence of the adopted models.

Despite several benefits provided to consumers [27] and the increasing adherence to the services [7], security issues of the cloud paradigm stand as a major challenge for its large-scale adoption [10]. According to Bouchenak [4], consumers aim at having at least equivalency between the security provided by the cloud and the one experienced on local environments. Besides, Huang [16] advocates that a joint effort between academia and industry is needed to solve the pressing security issues in cloud.

SLA corroborates the shared management vision provided by cloud deployment and service models, as the cloud provider must ensure customer requirements for the application, data, and infrastructure offered according to SLA definitions. This shared management is considered a security issue related to cloud management (including governance, compliance and legal issues) [13] and specifications and definitions of security requirements for cloud computing SLAs are still in embryonic stages [12] as it has only recently been considered a necessity [4]. Luna [24] advocates for the need of cloud management architectures based on security SLA requirements.

In order to face the open challenge of cloud management architectures based on security SLA requirements, this paper proposes a framework to orchestrate the lifecycle of security SLAs for cloud computing. The framework orchestrates the provisioning of cloud services and security mechanisms based on the security requirements defined by a SLA in an automated manner by following the four phases that compose the SLA lifecycle proposed by Rojas [31]. Furthermore, we present how the framework can be transparently integrated to cloud solutions such as OpenStack and how it supports the phases of the SLA lifecycle.

This paper is organized as follows. Sect. 2 presents the concepts related to the security issues of SLA agreements for cloud computing domains, the phases of an unified SLA lifecycle for cloud computing as well as the security requirements in SLA for cloud computing. It also presents related works to security SLA for the cloud computing context and gives a comparative analysis between the SLA lifecycle proposed and the related work. Sect. 3 presents the framework to orchestrate the lifecycle of a security SLA by contemplating the components description and their relations, internally and externally. In addition, the integration with the cloud solution OpenStack and the SLA lifecycle are presented. Also, a validation based on functional requirements is presented. Finally, Sect. 4 presents the conclusion and future work.

2 Service Level Agreement

This section presents the need to address the security issues of SLA contexts for cloud computing domains. Also, the phases of the SLA lifecycle for managing the need to define the level of services, which is required by the consumer, are presented also. Furthermore, presents the related works to security SLA for the cloud computing context and comparative analysis between the SLA lifecycle and the related works.

2.1 Security Issues

Despite the security issues of cloud computing being considered similar to the ones presented in ICT domains [25, 37], the need to address security in SLA for the cloud context was only recently proposed [4].

The current scenario of practices, obligations, recommendations and benefits related to addressing the security requirements in the SLA by cloud providers and consumers was surveyed by Rojas [31]. It was verified that the security aspects have been neglected in SLA contracts regarding the requirements specifications and its associated metrics. Furthermore, cloud providers do not have defined processes for managing the security requirements defined in the SLA and staff are not qualified for assisting and supporting customers during its definition.

Through the survey, the following challenges in cloud security SLA were identified: the architecture for managing security SLAs [4, 12], the definition of quantitative security metrics (and not just qualitative ones) [23], security SLA representation and security service disclosure [11, 26].

2.2 SLA Lifecycle

SLA lifecycle is composed of phases that are required to achieve SLA. Each phase supports the specific needs of providers and consumers. Twenty-one proposals for SLA lifecycle within the cloud computing context were found in the literature, with each proposal having between two and eight phases. These proposals were analyzed and unified based on their description, producing a taxonomy for the SLA lifecycle. Figure 1 illustrates the taxonomy and unified phases.

The SLA lifecycle is composed of four phases and subtopics of interest. The explanation of what comprises each phase is presented:

- Phase 1—Definition and Specification: This phase is characterized by the specification of requirements that would compose the SLA, as well as the identification of consumer needs and characteristics of service model adopted and supported by the service provider. This phase is covered in depth by the TM Forum [8]. It is composed by one or more activities: product/service development, initial speci-

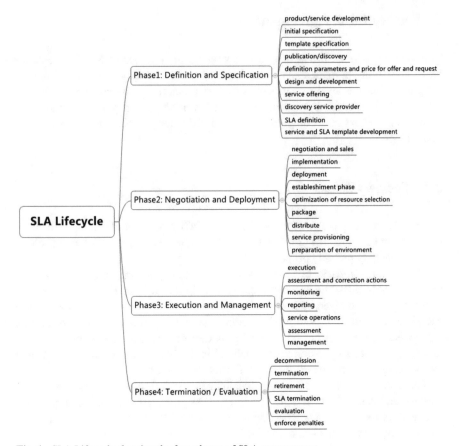

Fig. 1 SLA Lifecycle showing the four phases of SLA management

fication, template specification, publication/discovery, definition parameters and
price for offer and request, design and development, service offering, discovery
service provider, define SLA, service and SLA template development. The responsibility in this phase is shared between the service provider and consumer for both
definition and specification.

- Phase 2—Negotiation and Deploy: At this phase financial condition and acceptable levels of service defined between the service provider and the consumer
  are negotiated. Moreover, sanctions are defined for both parties in case of non-
  compliance in some agreed clauses. Also frequency and content of reports to be
  delivered by the service provider are defined. Then environment is provisioned and
  configured to deliver to consumers the requirements specified in the SLA. This
  phase is composed by one or more activities: negotiation and sales, implementation, deployment, establishment phase, optimization of resource selection, package, distribute, service provision and preparation of environment. In this phase
  negotiation task responsibility is shared between the service provider and consumer. Service provider is the sole responsible by the deployment task.

- Phase 3—Execution and Management: This phase is composed of services that are up and running in compliance with the requirements specified in the SLA. Concurrently with the execution, the following tasks are executed in the environment: real-time monitoring of the running instance, management of compliance requirements, emission of control reports, policy enforcement, corrective actions, reactive actions to be adopted and violations control. This phase is composed by one or more activities: execution, assessment and correction actions, monitoring, reporting, service operations, assessment, management. The service provider is the sole responsible for this phase.
- Phase 4—Termination/Evaluation: At this phase the termination of service due to the end of the contract, violation of the contract or user request. The instance is disabled in addition to the release of allocated resources and revocation of user access and grants. Issues related to billing of resources consumed and invoices issuing are treated. In the case of penalties due to non-compliance with any requirement by the service provider side, consumer can be discounted or be offered other compensations. This phase is composed by one or more activities: decommission, termination, retire, terminate SLA, evaluation, enforce penalties. In this phase the service provider is responsible for the termination task and the consumer is responsible for the evaluation task.

The phases of the SLA lifecycle assist the management concerns of cloud computing: the understanding of phases and required controls allows a comprehensive management of the shared responsibilities by consumers and cloud providers. Furthermore, the application of the SLA lifecycle for managing security concerns is fundamental for the adoption and evolution of the cloud paradigm.

2.3 SLA Security Requirements

Specifying security requirements in SLA for cloud computing services is a real need to consumers and service providers. It is necessary to understand how consumers, service providers and other stakeholders are dealing with the security issues in cloud computing and how these issues are addressed in the SLA contexts.

2.3.1 Service Provider Practices (PSP)

The following practices were identified regarding how to deal with security requirements in SLA cloud computing; by the service providers [1, 23, 25]:

- PSP 1. The service provider is solely responsible for determining the occurrence of security SLA violations.
- PSP 2. The SLA is composed by generic statements that inform how the provider must protect user data without specifying the level of protection and how this protection will be handled.

- PSP 3. Generic security measurements are made and stored by the service provider.
- PSP 4. There is no mandatory mention of how security measurements are performed by the provider.
- PSP 5. The absence of definitions and obligations to consumer data security.
- PSP 6. There is no specification of security service levels in the contract, thus preventing the user to make relevant decisions on your environment security.

Through these practices it is clear that the security aspects have been neglected by the service providers regarding the specification, monitoring of security requirements and associated metrics. It also points out the unpreparedness of consumers in dealing with security issues.

To help both providers and consumers deal with these issues, service providers associated with the CSA (Cloud Security Alliance) [28] created a public database called STAR (STAR—Security, Trust & Assurance Registry) that contains common security controls adopted by service providers.

2.3.2 Service Provider Obligations (OSP)

The obligations of the service provider related to security requirements [15, 17, 18, 25, 29, 33] include:

- OSP 1. Consider the adoption of good practices related to security and privacy. The security and privacy must be explicit, separate and clearly identified in the SLA document.
- OSP 2. When the service provider undergoes an security attack that causes damages to user data, it is the obligation of the service provider to restore data from the backup.
- OSP 3. Service provider must have a security professional to assist the consumer in the evaluation and selection of appropriate security mechanisms and parameters.
- OSP 4. Service provider should have a security incident response team (CSIRT).
- OSP 5. The confidentiality of consumer data must be taken into account. The consumer should be informed how the provider handles the confidentiality of data as well as related vulnerabilities and mitigation actions.
- OSP 6. Service provider shall evaluate and report security vulnerabilities, controls, services and mechanisms employed.
- OSP 7. When data is transferred to cloud, the provider, or whoever has the custody of the data, is the responsible for protecting and ensuring data security.
- OSP 8. Service provider must have relevant certifications regarding security standards (e.g. ISO 27001, PCI-DSS, HIPPA, etc.).

It is possible to verify that service provider needs to develop a process to deal with issues related to security and management. Special attention is recommended to support and assist customers during the entire definition and SLA deployment process.

2.3.3 Service Provider and Consumer Recommendation (RPC)

General recommendations to service provider and consumer [6, 9, 14, 17, 21, 22, 24, 25, 38] include:

- RPC 1. The SLA must specify security responsibilities and also includes disclosure of security vulnerabilities to consumers.
- RPC 2. Security requirements specified and agreed by SLA should be extended to all parties involved at any stage of the service deployment and management.
- RPC 3. Metrics and standards to evaluate the performance and effectiveness of information security management should be established and specified in the SLA.
- RPC 4. Data privacy policies must be included in the SLA by the service provider. Examples of policies: data processing, data storage and usage, during communications and data retention.
- RPC 5. The right to audit the service provider should be an SLA clause, enabling traceability and transparency to security requirements defined in the SLA.
- RPC 6. SLA should be composed of measurable security parameters.
- RPC 7. Customers should receive reports of your incidents in the environment and their evaluations.
- RPC 8. Financial compensation must be planned in case of failure for meeting the contracted services.
- RPC 9. Details about implementation of encryption algorithms should be specified.
- RPC 10. Access control policies should be specified.
- RPC 11. SLA may contain no quantitative parameters, such as: regulations, geographical restrictions for processing and or data store, also standards for business process (e.g. ISO 9001:2000).
- RPC 12. The security monitoring process including evidence collection must include from where the evidence is collected and who is responsible for collecting it.
- RPC 13. SLA should contain details on how security is maintained by the provider, its applied methods and how customer complaints are meet.
- RPC 14. Security settings in SLA should address the activities of operational and administrative management.
- RPC 15. The inclusion of security requirements in the SLA explicitly requires stakeholders to treat security issues.

These recommendations presented the indispensable need to address security requirements definitions in SLA, thus covering from the definition of parameters to be monitored to the definition of penalties in case of failure. This set of recommendations represent major challenges for service provider as well as for the involved stakeholders.

2.4 Security Requirements Versus SLA Lifecycle

Table 1 presents relations between the security requirements and proposed SLA life-cycle.

In Table 1 we can see 66% of the requirements identified by this research are related to Phase 1. This percentage matches the necessity of addressing security requirements in the SLA definition and specification and represents security concern in cloud computing and the necessity to control security issues to guarantee managed deployment of consumers services. The responsibility to achieve success in this fundamental phase is shared by both service provider and consumer, because for both of them it is important to define security requirements precisely. Despite the major percentage of this phase the existing security requirements are not well defined to help consumer or service provider to define quantifiable metrics or measurable security services. It requires more research and development to support this area with well defined security requirements, metrics and measurable security services.

The Phase 2 has 3% of the requirements as shown in Table 1 that is related to security services levels negotiation and environment deployment requested by the consumer. The responsibility of negotiation is shared by both service provider and consumer. The responsibility to properly deploy service requested by the consumer is unique by the service provider. It is important for the consumer to negotiate rights to audit the environment and financial compensations related to faults by the service provider. This phase is closely related to specifications made at Phase 1.

Phase 3, with 28% is related to execution and management of deployed service by the service provider. This phase is of entire responsibility of the service provider who manages the deployed services, supports adequately the service level of security requirements and measurable security services. In this phase cloud infrastructure needs to support security requirements and measurable security services. Therefore, cloud architecture has to be able to monitor, enforce security definitions and audit environment. The success in this phase is closely related to the Phase 1 when definitions of security requirements are defined and specified.

Phase 4 related to the termination of service by the service provider and evaluation of the service by the consumer has 3% of requirements as shown in Table 1. Just the financial compensation was considered in this phase. The low percentage of security requirements observed in both, this Phase and Phase 2 emphasizes the necessity of more research in these stages of SLA and development in security requirements for cloud computing.

2.5 Related Work

The necessity for addressing security requirements in cloud computing SLAs has become a research challenge for the academic community, standard institutes and enterprises. In this context eight related works were identified and analyzed in respect to SLA lifecycle.

Table 1 Relationship between SLA lifecycle and security requirements for cloud computing

	Phase 1 Definition and Specification	QTY	Phase 2 Negotiation and Deploy	QTY	Phase 3 Execution and Management	QTY	Phase 4 Termination / Evaluation	QTY
Service provider practices	PSP2, PSP5, PSP6	3			PSP1, PSP3, PSP4	3		
Service provider obligations	OSP1, OSP3, OSP5, OSP7, OSP8	5			OSP2, OSP4, OSP6	3		
Service provider and con-sumer recommendation	RPC1, RPC3, RPC4, RPC5, RPC6, RPC9, RPC10, RPC11, RPC13, RPC14, RPC15	11	RPC2	1	RPC7, RPC12	2	RPC8	1
Total	19		1		8		1	
Percentage	66%		3%		28%		3%	
Responsibility	Service provider (D, E), consumer (D, E)		Service provider (N, D), consumer (N)		Service provider (E, M), consumer ()		Service provider (T), consumer (E)	

- Garcia [12] proposed a framework to develop security metrics for evaluations by cloud providers, taking into account cloud models of service and deployment. This work contributes to Phase 1 of the SLA lifecycle because it addresses the issues related to the definitions and specifications of the security parameters.
- Luna [23] presents a set of metrics used to quantitatively compare security SLAs in cloud contexts. The CSA information base called STAR (Security, Trust & Assurance Registry) was a case study to compare the security practices applied by different cloud providers. Therefore, contributing to Phase 1 with security metrics that can be applied in SLAs. Also, it can be used to discover service providers and service offerings that are related to security features applied to Phase 1. In addition, it may be employed to assist negotiations (Phase 2), since enables customers to evaluate and compare the security features being offered by cloud providers.
- Luna [24] presents a quantitative and qualitative method to compare security requirements defined in the security SLA between cloud providers and consumers. This method is based on the QPT (Quantitative Policy Trees), an approach that enables the evaluation of security requirements by categories. This approach facilitates the comparison between security offers and service providers. The work contributes to Phase 1 due to the possibility of comparing security features by applying quantitative and qualitative metrics. In addition, it also contributes to Phase 2, since comparative results can be used in the negotiation process.
- Ullah [36] presents an automated tool to evaluate compliance levels related to security requirements adopted by cloud providers. This work supports Phase 1 because it identifies the security services offered by cloud providers, the security mechanisms applied and the related parameters.
- Silva [34] proposes a method to evaluate the provisioning of cloud resources based on hierarchical security metrics. The hierarchical metrics are developed applying the GQM (Goal-Question-Metric) approach. Thus contributing to Phase 2 due to the capacity of service provision based on security requirements and to Phase 1 due to generated security metrics that can be specified in SLAs.
- Rank [30] proposed an approach to deploy cloud resources based on the user roles and security policies defined in the SLA. The work contributes to Phase 2 due to the deployment of resources based on security requirements defined in the SLAs.
- Jegou [19] presented the VEP (Virtual Execution Platform) a component responsible for provisioning IaaS resources based on the requirements defined in the SLA. Furthermore, VEP is responsible for monitoring the instance deployed during the entire lifecycle. The work contributes to Phase 2 due to deployment resources based on security SLA and to Phase 3 due to the monitoring of activity during the execution of the deployed instance.
- Ferreira [11] proposes an architecture for monitoring security based on SLA for the IaaS service model. The monitoring approach avoids the necessity of installing a monitoring agent on the host. The work can support the Phase 3 due to the process for monitoring the execution of the deployed resources. Also, it contributes to Phase 2 due to deployment of resources based on security SLAs, and to Phase 1 due to an approach for representing security policies in SLAs.

Table 2 Relation between related works and SLA phases

Related work	Phase 1	Phase 2	Phase 3	Phase 4
Garcia [8]	1			
Luna [13]	1	2		
Luna [9]	1	2		
Ullah [16]	1			
Silva [17]	2	1		
Rank [18]		1		
Jegou [19]		1	2	
Ferreira [15]	3	2	1	
Total	6	6	2	0

Table 2 summarize the relation between the related works found in the literature and the SLA lifecycle phases.

It can be verified that Phase 1 is the most addressed, being the focus of four works and related in three others. Meaning that to define and to specify security metrics and security parameters is important and fundamental for the security management of cloud solutions. Methods to define this metrics and parameters are equally important. Phase 2 is the second most addressed with three related works focusing on it and other three related indirectly. The provisioning and deployment of cloud resources based on security metrics were the challenges faced by these works. Phase 3 has one related work to it directly and another related indirectly. The work is focused on the necessity for monitoring security. It can be verified that more development of this phase is needed in terms of mechanisms and methods for managing security in cloud environments, along with monitoring and auditing tools. Phase 4 has no contribution. It can be considered a consequence related to the necessity of evolving the other phases. However, it does not justify the lack of related works despite the issues appointed. In this case it can be considered a fertile area of future works. From the results of our analysis it can be verified the necessity of more research in the security SLA area and solutions related to Phases 3 and 4 due to their importance in the security context. Furthermore, the results corroborate the necessity of improving security in cloud environments as appointed by consumers and providers.

Based on related works it can be verified that there are no proposals that encompass the four phases of the lifecycle, just proposals to a specific phase. To meet this need a framework to orchestrate the security SLA lifecycle phases in an automated and integrated manner is proposed, as advocated by Bouchenak [4] and Garcia [12]. The proposed framework aims to fill this gap related to cloud management based on security requirements advocated by the academia and appointed as a necessity by cloud consumers and providers.

3   Security SLA Framework

This section presents the details of framework proposed by Rojas [32] and how it supports the phases of the SLA lifecycle, also their integration with the cloud solution
adopted. By analyzing the related works, it was verified the necessity for solutions to
orchestrate the lifecycle of security SLAs. In this context, a framework is proposed
to orchestrate the security requirements addressed in the SLA lifecycle in an automated and integrated manner, also solutions for individually supporting each phase
of the SLA lifecycle are presented.

The framework is composed of two sides, user and cloud provider. The user side
has available interfaces for communicating with the cloud provider side. The cloud
provider side comprises the proposed framework and their integration with the cloud
solution infrastructure. Figure 2 illustrates the framework

The user side provides the interface to allow users to access the cloud side and
define the security SLA required. It is composed of three communication interfaces
to support the interactions between both sides:

- SLA Interface: It is the GUI used to specify the security parameters and mechanisms to be applied by the cloud provider for the deployment the service requested.
  This interface could be integrated with the Client Application.
- SLA Client: It was designed to be a command line tool for testing the SLA engine
  and their integration with other functions during the framework development. At
  the end, it can also be used as an auditing tool by both user and service provider,
  or even an independent auditor.
- Client Application: It is the primary interface (GUI and CLI) provided by the
  cloud solution to the user (e.g. Horizon the web-based management interface for
  OpenStack).

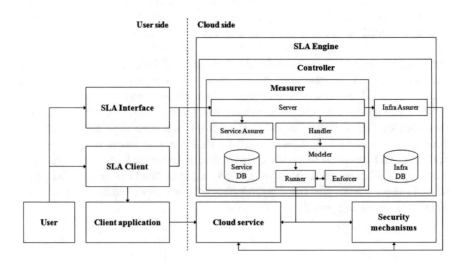

Fig. 2   Security SLA framework proposed for cloud computing

The cloud provider side contains the engine for orchestrating the management of the lifecycle of cloud services based on a SLA specified according with user, cloud infrastructure and security mechanisms related. The SLA Engine is composed of nine modules:

- Controller: responsible for managing the execution of modules. Ensures that the correct order of operations is performed and provides access to the database (Infra DB).
- Measurer: is the interface for writing and reading information related to the operations performed and accountability. It also provides access to the database (Service DB).
- Server: is the server side SLA Interface and SLA Client. Listens for incoming requests and updates information related to operations performed by the SLA Engine.
- Handler: Manages the incoming requests related to services, parameters and security mechanisms to be applied in the deployment of the service.
- Modeler: Translates the input parameters, related to services and security mechanisms, into adequate data structure for managing the deployment of service. It can be customized to generate the proper data structure in order to support the cloud solution adopted (e.g. OpenStack, OpenNebula, Eucalyptus, CloudStack, etc.). Then, more than one cloud solution could be adopted.
- Runner: Dispatch run orders, executing the adequate programs, sequences and security mechanisms. The executing task could also support distinct cloud solutions.
- Enforcer: Enforces the right commands and execution order (workflow). The order of execution and parameters are hashed and stored to support compliance and audit tasks.
- Service Assurer: Assures that services are being provided according to the specified security SLA and following the defined workflow. In addition, control information is stored in the database (Service DB). The Service DB stores the information about the service usage (requests received, operations performed, information assurance, user data). Moreover, the information stored is hashed to guarantee its integrity. Nonetheless, it is a data source that can support audit tasks.
- Infra Assurer: Collect information related to services and security mechanisms provided by the cloud available through the SLA Interface. It provides assurance on assessments of the hardware and software pieces regarding trust and correct operation. If the cloud infrastructure has TPM (Trusted Platform Module) and vTPM (Virtual Trusted Platform Module) resources, their primitives can be used to improve the assurance process [2]. In addition, control information is stored in the database (Infra DB). The Infra DB stores information related to the engine operations and cloud infrastructure resources (services, security mechanisms, trusted hardware).

The cloud provider side supplies cloud solutions, services and security mechanisms (SecMecs). The SecMecs can be native mechanisms related to the services provided or external designs provided by third parties integrating with the cloud solution. The current integration process between the cloud resources and framework are not automated. The automatic discovery of resources and their integrating to framework is a challenge that needs research.

The framework proposed here is independent from the cloud solution adopted for providing services and security mechanisms to users. The modules Modeler, Runner and Infra Assurer can support this feature. Besides, the framework can orchestrate cloud services and security mechanisms provided by more than one cloud solution

3.1 Lifecycle Integration

The framework was designed to support the SLA lifecycle phases. The relation between the lifecycle phases and framework modules, along with the lifecycle workflow is illustrated in the Fig. 3.

It can be verified that the Definition and Specification phase support the customer activities, to compose a contract for the desired SLA, which has to be established and accepted by the both parties, service provider and consumer. This activity is supported by the framework modules the SLA Interface, Server and Infra Assurer. After defining the SLA, it has to be Negotiated and then Deployed in the cloud solution. For the Negotiation activity the modules applied by previous phase are referred to. For the Deployment activity the framework provides the modules Runner, Enforcer and Service Assurer. After the deployment of the services and security mechanisms related, the requested service is applied. The adequate deployed environment has to be Executed and Managed to ensure that the security levels defined are attended to. To support the phase activities the framework provides the modules Runner, Enforcer and Service Assurer. For the service release instances, there is a Termination phase followed by Evaluation related to accounting and billing. To support Termination the framework provides the modules Runner, Enforcer and Service Assurer. To support Evaluation the framework uses the module Service Assurer. Considering the

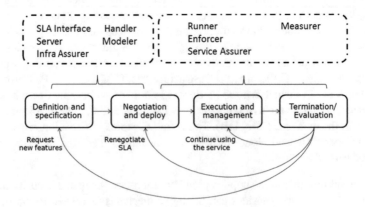

Fig. 3 Relation between SLA lifecycle and proposed framework

workflow of a SLA lifecycle, it is possible to migrate from the last phase back to the first phase (e.g., if the customer requests new features or if the contract has to be changed). It is also possible migrate back to the second phase, if the SLAs have to be renegotiated. Finally, it is possible migrate back to the third phase if the customer simply wants to continue using the service with the previously set SLA definitions.

3.2 OpenStack Integration

To support the proposed framework, a cloud solution is necessary in order to provide services and the security mechanisms related. The integration process between the cloud solutions require open access to information related to internal architecture and source code of services provided (e.g., virtual machines, storage, images, etc.). Each service possesses its internal communication and security mechanisms, and was identified through source code analysis. The OpenStack Icehouse release series was nominated as cloud solution to support this integration due to their importance in the cloud computing scenery, for providing the necessary resources and for its open source code. Furthermore, OpenStack is considered the largest most active community when compared to solutions such as OpenNebula, Eucalyptus and Cloud-Stack [20].

The Nova service source code, which provides compute resources (VM) to users, was analyzed in relation to their existing security mechanisms. Two functionalities and four security mechanisms are going to be presented. They are examples of native, non-native, and proposal of new mechanisms. The security mechanisms presented had their impact analyzed in terms of the CIA triad [3] based on the NIST security standards [5, 35]. Figure 4 illustrates the mechanisms and their security analysis.

Three functions of the Boot VM functionality can be improved. The first is hash VM related files, this action increases the service integrity. This is an example of native mechanism. The second checks the reliability of the physical machine (TPM-based infra), this action increases both integrity and availability. The third seals the VM by using TPM or vTPM (VM assurance), this action increases service confidentiality and integrity. The cloud infrastructure may have physical computing resources TPM-based for deployment to customers and vTPM to support the VM deployment. The TPM primitives (e.g. attestation, binding, sealing, etc.) can support the non-native second and third functions. The Delete VM functionality can be developed

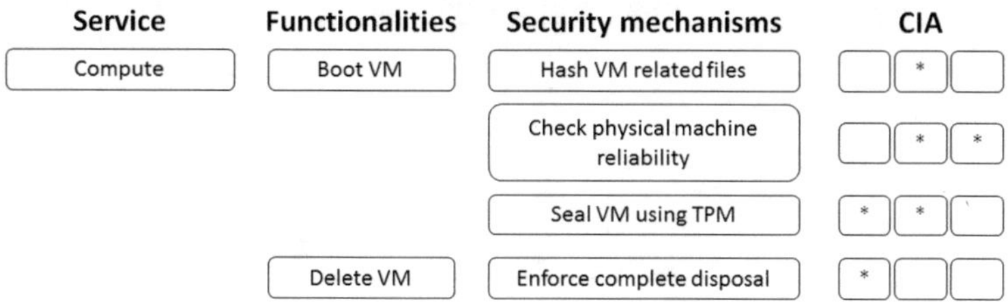

Fig. 4 Security mechanism for Nova service and their analysis

to enforce the complete disposal process, this action increases service confidentiality. This is an example of security mechanism that can be developed by the cloud provider or third parties.

The same analysis can be made to other OpenStack services related to security mechanisms and their security impacts. The security mechanisms can be applied by the framework throughout the entire SLA lifecycle, supporting the need for management and enforcement of contracted security requirements.

To present the integration of the framework proposed (SLA Interface and SLA Engine) within cloud service (Nova) and security mechanism (TPM-based infra) the sequence diagrams were elaborated. The cloud operations (define SLA and deploy VM) and security mechanisms (TPM-based infrastructure and VM assurance) were selected to produce the sequence diagrams that illustrate the modules operations. The framework side of the SLA Interface defines a SLA and selects the SLA Engine.

3.2.1 Sequence Diagram: Define SLA

The process of SLA definition is illustrated in Fig. 5.

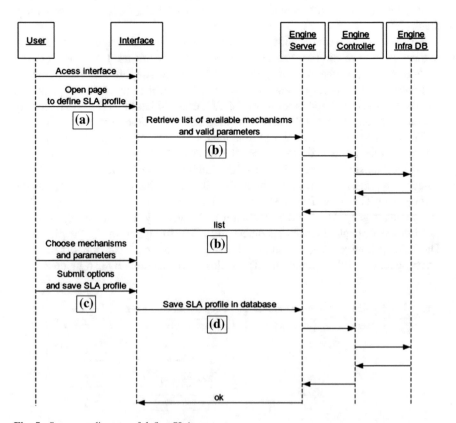

Fig. 5 Sequence diagram of define SLA process

The process is illustrated by the following steps: (a) user accesses the SLA Interface (GUI) and opens the page for defining the SLA. (b) The interface retrieves the information to the user from the SLA Engine by presenting the available security mechanisms related to the Nova service. The user then sets the desired SLA definitions. (c) The set configuration of the security mechanisms and parameters related are submitted by the user. (d) The SLA is configured as a profile which is stored in the Engines Service DB and the provider confirms the operation to the user.

3.2.2 Sequence Diagram: Deploy a VM

The VM deployment process is presented in two phases. The first is a definition of the features related to VM is illustrated in Fig. 6.

The process is illustrated by the following steps: (a) the user accesses the OpenStack Dashboard and open the create VM option. (b) The interface retrieves information from the SLA profile. (c) The SLA Engine queries the Service DB and sends the defined and available profiles. (d) The user sets up the VM parameters (flavor, memory, image, etc.) and the desired SLA profile. Finally, the user submits the request to OpenStack. In the second phase the VM parameters and SLA profile submitted are handled by the SLA Engine. This phase is illustrated in Fig. 7.

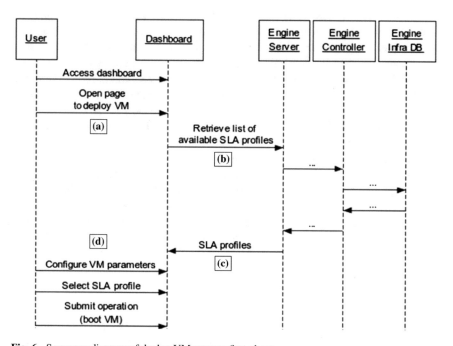

Fig. 6 Sequence diagram of deploy VM process first phase

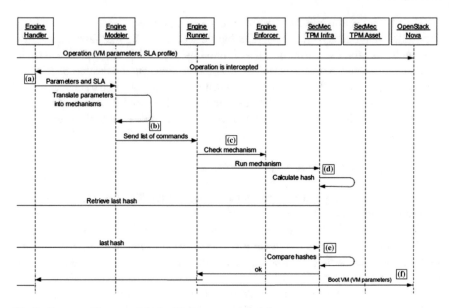

Fig. 7 Sequence diagram of deploy VM process second phase

The process is illustrated by the following steps: (a) the request is intercepted by Handler, which reads the SLA profile information and sends it to Modeler. (b) Modeler receives the parameter of the SLA profile and defines the adequate commands; then, in the correct order, sends it to be run by Runner. (c) Runner executes the commands while checking with the Enforcer, which guarantee that the right commands are being executed in the right order (in compliance with the SLA). (d) Runner executes the command which gathers the information of infrastructure (hardware and software) and hashes it. The information is processed by the TPM module and stored in Infra DB. This hashed information is also compared to the previously stored database in order to verify modification occurrences of the infrastructure. (e) After confirmation, Runner executes the command to Boot VM with the requested parameters.

Through the sequence diagrams are presented how the user defined the SLA and how this SLA is applied within the Nova cloud services in order to deploy a VM with security features provided by the mechanism. Then, the communication and interaction between the framework modules and cloud solution are presented.

3.3 Preliminary Validation

The proposed framework was designed based on functional (FR) and non-functional (NFR) requirements defined by the team members attending the necessity to orchestrate the deployment of security SLAs for the cloud environment. The validations of requirements (VFR/VNFR) are presented:

- FR1: The solution must provide a graphical interface to allow easy interaction between the user and the mechanisms, providing visualization tools to define the required SLA levels.
- VFR1: To satisfy the requirement the SLA Interface and SLA Client was designed, providing the required interface.
- NFR2: The solution must provide a command line interface to allow easy integration between itself and other programs or scripts.
- VNFR2: To satisfy the requirement the SLA Client was designed, providing the required interface.
- NFR3: The solution must provide a transparent communication interface to interact with any type of interface (e.g. via HTTP requests).
- VNFR3: This requirement was satisfied by the Server module that provides a transparent communication interface.
- FR4: The solution must handle the input parameters defined by the user, thus defining which security mechanisms and cloud services are necessary to provide the SLA levels specified by the user.
- VFR4: To satisfy the requirement the Handler was designed, providing the adequate services and mechanisms to be applied in the deployment of the service.
- FR5: The solution must allow integration to multiple cloud environments (e.g. OpenStack, OpenNebula, AWS), without being restricted to any specific environment.
- VFR5: This requirement was satisfied by the Modeler module that provides a translate mechanisms to the adequate cloud solution.
- FR6: The solution must define the execution order of commands while providing the SLA levels specified.
- VFR6: To satisfy the requirement the Runner was designed, providing the interface to execute the right commands to the cloud solution adopted.
- FR7: The solution must provide assurance regarding both the execution order and the SLA levels actually delivered, and the assurance must cover both how the service is delivered and how the infrastructure is deployed.
- VFR7: This requirement was satisfied by the Service Assurer and Infra Assurer modules that provide the adequate mechanism to guarantee that execution order and service comply with SLA agreed. Through these modules is possible to audit the environment, verifying that the SLA was complied by the cloud provider.
- FR8: The solution must provide accountability related to its own operations and also the services and actions performed in the cloud, and the information should be stored in adequate databases.
- VFR8: To satisfy the requirement the Service Assurer and Infra Assurer modules was designed, each module store the operations performed by the cloud in the respective database, providing the accountability of the environment.

It can be verified that the proposed framework can support the definition of security requirements through SLA and orchestrate the deployment of cloud service and security mechanism based no defined SLA. Beyond that the framework can provide the auditability, which is an interesting and requested feature for cloud providers and consumers.

4 Conclusion and Future Work

This work presents the need for addressing security requirements in cloud computing SLA as a fundamental aspect of development, consolidation, advancement and management of cloud technology. The process defined by the SLA lifecycle can assist in the management of cloud services due to the shared responsibilities between providers and consumers required by the cloud environment. This management is accomplished by well-defined phases and it can also improve the control and mitigation of security issues desired for this environment, thereby reducing the risks from both sides. A proposed framework was presented and validated as a solution to manage the lifecycle of security SLAs for the cloud context in an automated and flexible manner, improving the security and compliance required by consumers. The outcome of the analysis advocates that security research requirements in cloud computing are currently at embryonic stages, therefore, efforts from the security communities to properly define those security requirements for cloud environments, as well as the metrics to properly measure cloud services, are needed in order to improve the desired management of security issues required by consumers and cloud providers. For future work, it is necessary to fully implement the framework, test and evaluate the results, along with their integration with OpenStack. Furthermore, automated discovery and integration of security mechanisms are interesting topics for future research.

Acknowledgements This work was supported by the Innovation Center, Ericsson Telecomunicações S.A., Brazil.

References

1. Baudoin, C., Flynn, J., McDonald, J., Meegan, J., Salsburg, M., Woodward, S.: Public cloud service agreements: what to expect and what to negotiate. Technical report, Cloud Standards Customer Council (CSCC) (2013). http://www.cloud-council.org/publiccloudSLA.pdf
2. Berger, S., Cáceres, R., Goldman, K.A., Perez, R., Sailer, R., van Doorn, L.: vtpm: virtualizing the trusted platform module. In: Proceedings of the 15th Conference on USENIX Security Symposium, vol. 15, USENIX-SS'06. USENIX Association, Berkeley, CA, USA (2006). http://dl.acm.org.ez67.periodicos.capes.gov.br/citation.cfm?id=1267336.1267357
3. Bishop, M.A.: Computer Security: Art and Science. Addison-Wesley Professional (2002)
4. Bouchenak, S., Chockler, G., Chockler, H., Gheorghe, G., Santos, N., Shraer, A.: Verifying cloud services: present and future. SIGOPS Oper. Syst. Rev. **47**(2), 6–19 (2013). doi:10.1145/2506164.2506167
5. Bowen, P., Hash, J., Wilson, A.: Information security handbook: a guide for managers. Technical Report 800-100, National Institute of Standards and Technology (NIST) (2006)
6. CPNI: Information security briefing 01/2010—cloud computing. Technical report, Centre for the Protection of National Infrastructure (CPNI) (2010). http://www.cpni.gov.uk/Documents/Publications/2010/2010007-ISB-cloud-computing.pdf
7. Csaplar, D.: Who is adopting the public cloud faster ? north america or europe ? Technical report, Aberdeen Group (2013). http://www.aberdeen.com/Aberdeen-Library/8565/AI-public-cloud-adoption.aspx

8. Damm, G., Bain, G., Timms, J., Philippart, L., Roman, R., Deheus, J., Cruz, A., Best, I., Milham, D., Alhakbani, M.A.: Gb917—sla management handbook version 3.0. Technical report, TeleManagement Forum (2012). http://www.dmtf.org/sites/default/files/standards/documents/DSP2029/20-1.0.0a.pdf
9. Dekker, M., Hogben, G.: Survey and analysis of security parameters in cloud slas across the european public sector. Technical report, ENISA—European Network and Information Security Agency (2011). http://www.enisa.europa.eu
10. Fernandes, D.A.B., Soares, L.F.B., Gomes, J.V., Freire, M.M., Inácio, P.R.M.: Security issues in cloud environments: a survey. Int. J. Inf. Secur. **13**(2), 113–170 (2014). doi:10.1007/s10207-013-0208-7
11. Ferreira, A.S.: Uma arquitetura para monitoramento de segurança baseada em acordos de níveis de serviço para nuvens de infraestrutura. Instituto de Computação, Universidade Estadual de Campinas, UNICAMP, Dissertação de mestrado (2013)
12. Garcia, J.L., Ghani, H., Germanus, D., Suri, N.: A security metrics framework for the cloud. In: Lopez, J., Samarati, P. (eds.) SECRYPT, pp. 245–250. SciTePress (2011). http://dblp.uni-trier.de/db/conf/secrypt/secrypt2011.html#LunaGGS11
13. Gonzalez, N.M., Miers, C., Redigolo, F.F., Carvalho, T.C.M.B., Jr., M.A.S., Nslund, M., Pourzandi, M.: A quantitative analysis of current security concerns and solutions for cloud computing. J. Cloud Comput. Adv. Syst. Appl. **11**(1), 1–18 (2012)
14. Henning, R.R.: Security service level agreements: quantifiable security for the enterprise? In: Kienzle, D.M., Zurbo, M.E., Greenwald, S.J., Serbau, C. (eds.) NSPW, pp. 54–60. ACM (1999)
15. Hogben, G., Dekker, M.: Procure secure: a guide to monitoring of security service levels in cloud contracts. Technical report, ENISA—European Network and Information Security Agency (2012). http://www.enisa.europa.eu
16. Huang, W., Ganjali, A., Kim, B.H., Oh, S., Lie, D.: The state of public infrastructure-as-a-service cloud security. ACM Comput. Surv. **47**(4), 68:1–68:31 (2015). doi:10.1145/2767181
17. ITU-T: Focus group on cloud computing technical report part 1. Technical report, ITU-T (2012). http://www.itu.int/en/ITU-T/focusgroups/cloud/Pages/default.aspx
18. Jaatun, M., Bernsmed, K., Undheim, A.: Security slas an idea whose time has come? In: Quirchmayr, G., Basl, J., You, I., Xu, L., Weippl, E. (eds.) Multidisciplinary Research and Practice for Information Systems. Lecture Notes in Computer Science, vol. 7465, pp. 123–130. Springer, Berlin Heidelberg (2012)
19. Jegou, Y., Harsh, P., Cascella, R., Dudouet, F., Morin, C.: Managing ovf applications under sla constraints on contrail virtual execution platform. Network and service management (CNSM). 2012 8th International Conference and 2012 Workshop on Systems Virtualiztion Management (svm), pp. 399–405. Las Vegas, NV (2012)
20. Jiang, Q.: Cy13-q4 community analysis openstack vs opennebula vs eucalyptus vs cloudstack. Technical report, IEEE Organization (2014). http://www.qyjohn.net/?p=3432
21. Kandukuri, B.R., Paturi, V., Rakshit, A.: Cloud security issues. In: IEEE International Conference on Services Computing, 2009. SCC '09, pp. 517–520 (2009)
22. Ken, R., Harris, D., Meegan, J., Pardee, B., Le Roux, Y., Dotson, C., Cohen, E., Edwards, M., Gershater, J.: Security for cloud computing: 10 steps to ensure sucess. Technical report, Cloud Standards Customer Council (CSCC) (2012). http://www.cloud-council.org/Security_for_Cloud_Computing_Final-080912.pdf
23. Luna, J., Ghani, H., Vateva, T., Suri, N.: Quantitative assessment of cloud security level agreements—a case study. In: In Proceedings of the International Conference on Security and Cryptography, SECRYPT 2012, pp. 64–73. SciTePress (2012). http://www.deeds.informatik.tu-darmstadt.de/fileadmin/user_upload/GROUP_DEEDS/Publications/conf/secLA_eval.pdf
24. Luna, J., Langenberg, R., Suri, N.: Benchmarking cloud security level agreements using quantitative policy trees. In: Proceedings of the 2012 ACM Workshop on Cloud Computing Security Workshop, CCSW '12, pp. 103–112. ACM, New York, NY, USA (2012). doi:10.1145/2381913.2381932
25. Meegan, J., Singh, G., Woodward, S., Venticinque, S., Rank, M., Harris, D., Murray, G., Di Mastirno, B., Le Roux, Y., McDonald, J., Kean, R., Edwards, M., Russel, D.,

Malekkos, G.: Practical guide to cloud service level agreement. Technical report, Cloud Standards Customer Council (CSCC) (2012). http://www.cloudstandardscustomercouncil.org/2012_Practical_Guide_to_Cloud_SLAs.pdf

26. Meland, P.H., Bernsmed, K., Jaatun, M.G., Undheim, A., Castejon, H.: Expressing cloud security requirements in deontic contract languages. In: Leymann, F., Ivanov, I., van Sinderen, M., Shan, T. (eds.) CLOSER, pp. 638–646. SciTePress (2012). http://dblp.uni-trier.de/db/conf/closer/closer2012.html#MelandBJUC12

27. Mell, P., Grance, T.: The nist definition of cloud computing. Technical Report 800-145, National Institute of Standards and Technology (NIST) (2011). http://csrc.nist.gov/publications/nistpubs/800-145/SP800-145.pdf

28. Online: Cloud security alliance—csa (2014). https://cloudsecurityalliance.org/

29. Patel, S.G., Jethava, G.B.: A review on sla and various approaches for efficient cloud service provider selection. Int. J. Eng. Res. Technol. **1**(1) (2012)

30. Rak, M., Liccardo, L., Aversa, R.: A sla-based interface for security management in cloud and grid integrations. In: 2011 7th International Conference on Information Assurance and Security (IAS), pp. 378–383 (2011)

31. Rojas, M.A.T., Gonzalez, N.M., Sbampato, F., Redigolo, F., de Brito Carvalho, T.C.M., Nguyen, K.K., Cheriet, M.: Inclusion of security requirements in sla lifecycle management for cloud computing. In: 2015 IEEE 2nd Workshop on Evolving Security and Privacy Requirements Engineering (ESPRE), pp. 7–12 (2015). doi:10.1109/ESPRE.2015.7330161

32. Rojas, M.A.T., Gonzalez, N.M., Sbampato, F.V., Redgolo, F.F., Carvalho, T., Ullah, K.W., Nslund, M., Ahmed, A.S.: A framework to orchestrate security sla lifecycle in cloud computing. In: 2016 11th Iberian Conference on Information Systems and Technologies (CISTI), pp. 1–7 (2016). doi:10.1109/CISTI.2016.7521372

33. Schnjakin, M., Alnemr, R., Meinel, C.: Contract-based cloud architecture. In: Proceedings of the Second International Workshop on Cloud Data Management, CloudDB '10, pp. 33–40. ACM, New York, NY, USA (2010). doi:10.1145/1871929.1871936

34. Silva, C.A.D., Ferreira, A.S., Geus, P.L.D.: A methodology for management of cloud computing using security criteria. In: 1st Latin American Conference on Cloud Computing and Communications (LatinCloud), pp. 49–54. IEEE, Porto Alegre, Brasil (2012)

35. Stoneburner, G.: Underlying technical models for information technology security. Technical Report 800-33, National Institute of Standards and Technology (NIST) (2001)

36. Ulla, K.W.: Automated Security Compliance Tool for the Cloud. Department of Telematics, Norwegian University of Science and Technology, NTNU, Master (2012)

37. Venters, W., Whitley, E.A.: A critical review of cloud computing: researching desires and realities. JIT **27**(3), 179–197 (2012)

38. Whiteside, F., Iorga, M., Badger, L., Mao, J., Chu, S.: Challenging security requirements for us government cloud computing adoption. Technical report, National Institute of Standards and Technology (NIST) (2012). http://www.nist.gov/customcf/get_pdf.cfm?pub-id=912695

Influence of User's Criteria Preferences for Open Source ITIL Tools Evaluation by Simple MCDM

Lukas Kralik, Roman Jasek and Petr Zacek

Abstract This paper responds to requirement to improve the orientation between offered SW, as ITIL tools. There are really a lot of offered tools and very often this fact may leads to the poor implementation of ITIL. Main objective of this paper is to provide an overview how users' preferences change in a short time and primarily the comparison between expert evaluation and evaluation with utilization of simple multi-criterial decision making methods. Simultaneously, the research described in this paper will serve for further work on creating a methodology for evaluation of ITIL tools.

1 Introduction

With the development of information and communication technologies (ICT) and their intrusion into all sectors, gaining management and delivery of IT services different dimension and meaning. The quality of providing or managing of IT services can greatly affect the operation or performance of the company. For this reason it was introduced, the now internationally acclaimed standard known as ITIL. It is an abbreviation for Information Technology Infrastructure Library. It is a set of concepts and practices that allow better planning, use and improve the use of IT, whether by the providers of IT services or by the customers.

ITIL is a collection of books in the form of extensive and widely available manual for IT service management. The experiences and recommendations have become best practices. Provide sufficient flexibility to adapt the recommendations from books ITIL requirements and needs of a specific corporation. ITIL provides a

L. Kralik (✉) · R. Jasek · P. Zacek
Tomas Bata University in Zlin, Zlin, Czech Republic
e-mail: kralik@fai.utb.cz

R. Jasek
e-mail: jasek@fai.utb.cz

P. Zacek
e-mail: zacek@fai.utb.cz

© Springer International Publishing AG 2018
Á. Rocha and L.P. Reis (eds.), *Developments and Advances in Intelligent Systems and Applications*, Studies in Computational Intelligence 718,
DOI 10.1007/978-3-319-58965-7_10

free available framework, covering the entire cycle of IT services. ITIL is suitable
for all companies that operate IT services. As a framework, ITIL is full of tips,
warnings, knowledge, omissions, instruction, warnings and things to do or not do.
One of the greatest benefits of ITIL is a fact that it is based on experience of others
[1–6].

According current version of ITIL v3, it is possible to say that ITIL tool is an
arbitrary software tool which use leads to provably improve and streamline the
providing and managing IT services. There is only one condition—it must be a SW
[7].

The uses of ITIL tools are complicated due to the wide range of offered tools and
often very expensive. This caused and to a certain extent still causes small and
medium companies are disinterest of the use of ITIL. On the other hand, recently is
beginning to discover significant amounts of Free and Open Source SW even
between ITIL tools [7, 8].

2 Methods

Three methods of multicriterial decision making were chose for the purpose of this
research.

1. Scoring method
2. Fuller's method (pairwise comparison)
3. Saaty's method (quantitative pairwise comparison)

The results of these methods were compared with an evaluation from users. The
main objective of this comparison is to design a methodology for evaluation ITIL
tools. The intended methodology must be simple and understandable for every
reader. This fact is a reason for the choice of above mentioned methods. Mathe-
matic operations in these methods are simple and it is clear to understand them
without any deeper knowledge from mathematics [8].

2.1 Scoring Method

This method assumes that user is able to quantitatively evaluate an importance of
criteria. A user evaluates i-th criterion with b_i value if this value is in selected scale
(e.g. $b_i \in < 0, 100 >$). A higher value means a higher importance of evaluated
criterion. A user can assign same value during the evaluation process and also user
does not have to choose only integers. Although, scoring method requires quantitative evaluation from user but at the same time it allows expression of subjective
preferences.

$$v_i = \frac{b_i}{\sum\limits_{i=1}^{k} b_i} \tag{1}$$

The final calculation of weights is simple. Assigned value for each weight is divided by the sum of all values (Eq. 1).

2.2 Fuller's Method

Also it is known as method of Fuller's triangle or mainly pairwise comparison (Table 1). This method exists in many modifications and it is determined for finding of preferential relations between pair of criteria. In the simplest modification of this method, the number of preferences is found out with the respect to all other criteria [9, 10]. This should be done according to Table 1. If criterion in row is more important than a criterion in column then the number 1 is typed into the cell otherwise 0. In agreement with the number of preferences, normalized weights are determined by the following Eq. (2) [11]

$$v_i = \frac{f_i}{m(m-1)/2} \tag{2}$$

f_i number of preferences of i-th criterion
m number of criteria
$m(m-1)/2$ number of comparisons

The disadvantage is a fact while some criterion has 0 preferences than its weight will be 0. That is a problem because this criterion does not be insignificant [7, 11]. This problem is solved by simple modification that respects indifference (same significant criteria). In this case, the cell is filled by the number 0.5 [10].

Table 1 Weights determination via pairwise comparison—fuller's method

Criterion	K_1	K_2	K_3	...	K_n
K_1		1	0	...	1
K_2			0	...	0
K_3				...	1
...				...	...
K_{n-1}					0
K_n					

2.3   Saaty's Method

This method is very often called as method of quantitative pairwise comparison (s_{ij}; $i, j = 1, 2, \ldots m$). The s_{ij} defines matrix S and its elements are interpreted as an estimation of division of weights (v_i and v_j) for i-th criterion and j-th criterion [9, 10].

$$s_{ij} \approx \frac{v_i}{v_j}, \, for \, i, j = 1, 2, \ldots, m. \tag{3}$$

Weights should be approximated from a condition that S matrix (s_{ij}) differ only minimally from V matrix (matrix contained from weights—v_{ij}):

$$D = \sum_{i=1}^{m} \sum_{j=1}^{m} \left[s_{ij} - \left(\frac{v_i}{v_j} \right)^2 \right] \to \min, \, \sum_{i=1}^{m} v_1 = 1 \tag{4}$$

A special vector of S matrix, which is assigned to the highest number, is used for approximation of weights. The Saaty's scale of relative importance is used for defining s_{ij} (described in more details in Table 2) [11].

It is possible to divide this method into 2 steps. The first one is very similar to pairwise comparison (Fuller's method). The next step is different because this method determines a size of preference with using the scale of relative importance [7, 11].

Determination of weights should be done by minimization method according to Eq. 4 [11]. The solution of S matrix helps to calculate weights v_i with the using of relative weight R_i (Eqs. 5, 6) [9, 10]

$$S_i = \prod_{j=1}^{m} s_{ij} \tag{5}$$

$$R_j = (S_i)^{1/m}, v_i = \frac{R_j}{\sum\limits_{i=1}^{m} R_j} \tag{6}$$

Table 2 Saaty's scale of relative importance

Intensity	Definition
1	Equally important
3	Slightly more important
5	More important
7	Strongly important
9	Absolutely more important
2, 4, 6, 8	intermediate values

3 Evaluation Criteria

Availability of Free and Open Source ITIL tools on the market is really wide that cause very difficult orientation between them. This problem is also related to selection of the most appropriate tools for a specific company. Therefore, below are defined and described the basic criteria for selection and evaluation of these tools [7]. However, it is important to say that each company may have different requirements and other criteria. So, same tool is useful for one company and at the same time might be useless for another company.

Basic criteria for Free and Open Source ITIL tools are divided into several groups (Table 3):

1. Product Functionality
2. Requirements for Free and Open Source Project
3. Specifications
4. User friendliness [8, 12, 13]

3.1 Product Functionality

Criteria relating to the functionality vary by application category. A large number of features do not necessarily mean that the application is better than competing product with a shorter list of features. This point cannot be assessed quantitatively as a measurable criterion of selection, but rather as an overview which may apprise readers and provide them information about the basic functions of the product [8, 12, 13].

This part mainly affects overall evaluation because users (respondents) from the survey provided a complex evaluation.

Table 3 List of criteria

Criterion	Abbreviation
Duration of the project	C1
Community	C2
Commercial support	C3
Documentation	C4
Trial/demo version	C5
HW requirements—RAM	C6
HW requirements—CPU	C7
Supported OS	C8
Integration	C9
Configuration	C10
User interface	C11
Language	C12

3.2 Requirements for Free and Open Source Project

Open source project is meant organizing and managing a group of people who are involved in the development of the product.

Primary criteria related to Open Source

- Duration of the project; version in which the product is available.
- License, under which the product is offered.
- Community
- Option of commercial support.
- Appropriate documentation—is the absence of the necessary documentation was in the selection of appropriate tools stumbling block relatively large number of projects. The basic requirement in this case, I consider the existence of technical documentation and user documentation.
- Demo application—trial version [8, 12, 13].

3.3 Specifications

Most of the Free and Open Source products use of ready-made programs usually also available under any other Free or Open Source licenses. This covers programs such as the Apache web server, or database servers MySQL, PostgreSQL, e-mail servers Postfix and so on [8, 12, 13].

Technical parameters are therefore a considerable amount and in particular, for each of this software may vary. Therefore, it is evaluation only directly influenced by the following parameters:

- HW requirements
- Supported operating systems—Cross-platform
- Integration with other SW
- Difficulty of configuration [8, 12, 13]

Other parameters such as licenses, programming language, etc. are given only as a parameter list and have only informative value to the end user, which can serve to more specific evaluation according to the requirements of the specific company [8, 12, 13].

3.4 User Friendliness

User friendliness is the main parameter that affects the user's ability to learn to work with a new product and use all functions. Improperly designed user interface can greatly influence user's work [8, 12, 13].

Evaluation of this criterion is very subjective and based primarily on practical experience. At the same time here enter localization—used language and of course the entire GUI (Graphic User Interface). Some tools are merely for the Command line [8, 12, 13].

4 Results

Selected MCDM methods were applied on 2 very spread Open Source ITIL tools—Zenos and Nagios.

Whole comparison contains few minor parts. The first one was collecting data from ca. 40 users (respondents)—simple survey. As was already said, users provided complex expert evaluation which was based on their practical experiences. So this evaluation included functionality which was omitted from the evaluation via MCDM methods.

The second step was another survey which was more complex. Second survey was focused on individual preferences for mentioned criteria. 10 participants had to think about which attribute (criterion) is more important for them than others. Firstly, they made pairwise comparison for all criteria. The number of comparisons is given by:

$$C_k = \frac{n!}{k!(n-k)!} \tag{7}$$

where n is equal to the number of criteria and k is 2 for pairwise comparisons. The resulting matrix is given on the basis of average preferences (Table 4).

Table 4 Pairwise comparison matrix

	C1	C2	C3	C4	C5	C6	C7	C8	C9	C10	C11	C12
C1		1	1	0.5	1	1	1	1	1	1	1	0.5
C2	0		0	0	0.5	0.5	0.5	0.5	0.5	0	0	0
C3	0	1		0	1	1	1	1	1	0.5	0	1
C4	0.5	1	1		1	1	1	1	1	1	1	1
C5	0	0.5	0	0		1	1	1	0.5	0	0	1
C6	0	0.5	0	0	0		0.5	0	0	1	0	0.5
C7	0	0.5	0	0	0	0.5		0	0	0.5	0	1
C8	0	0.5	0	0	0	1	1		1	1	0	0
C9	0	0.5	0	0	0.5	1	1	0		1	0.5	0
C10	0	1	0.5	0	1	0	0.5	0	0		0	0
C11	0	1	1	0	1	1	1	1	0.5	1		0
C12	0.5	1	0	0	0	0.5	0	1	1	1	1	

Table 5 Scoring method

Criterion	Average points	σ (+/-)
Documentation	90.4	3.9
Duration of the project	89.4	3.2
User interface	72.1	4.6
Commercial support	72.1	3.9
Trial/demo version	58.7	7.1
Configuration	52.9	9.1
Integration	50.8	5.5
Community	46.8	7.9
Supported OS	46.7	10.7
Language	46.1	8.7
HW requirements—CPU	33.2	11.5
HW requirements—RAM	27.6	8.6

Table 6 Matrix for quantitative pairwise comparison

	C1	C2	C3	C4	C5	C6	C7	C8	C9	C10	C11	C12
C1	1	7.00	3.00	1.00	6.00	5.00	8.00	4.00	5.00	3.00	5.00	1.00
C2	0.14	1	0.25	1.00	1.00	1.00	1.00	0.33	2.00	0.20	4.00	9.00
C3	0.33	4.00	1	0.25	6.00	7.00	7.00	3.00	2.00	1.00	0.50	8.00
C4	1.00	1.00	4.00	1	9.00	6.00	6.00	4.00	5.00	6.00	4.00	2.00
C5	0.17	1.00	0.17	0.11	1	5.00	6.00	2.00	1.00	0.11	8.00	4.00
C6	0.20	1.00	0.14	0.17	0.20	1	1.00	0.25	0.20	3.00	0.50	1.00
C7	0.12	1.00	0.14	0.17	0.17	1.00	1	0.33	0.20	1.00	0.17	3.00
C8	0.25	3.00	0.33	0.25	0.50	4.00	3.00	1	1.00	5.00	0.33	0.33
C9	0.20	0.50	0.50	0.20	1.00	5.00	5.00	1.00	1	7.00	1.00	0.50
C10	0.33	5.00	1.00	0.17	9.00	0.33	1.00	0.20	0.14	1	0.20	0.25
C11	0.20	0.25	2.00	0.25	0.12	2.00	6.00	3.00	1.00	5.00	1	0.25
C12	1.00	0.11	0.12	0.50	0.25	1.00	0.33	3.00	2.00	4.00	4.00	1

After that, participants had to assign points to every criterion according to its importance. They had a scale from 0 to 100 points where 100 is the most important criterion and 0 the less. It was possible to assign the same number of points for different criteria. This meant that these criteria was same important. Also, the overall sum did not have to be exactly 100 (Table 5).

Finally, they made another pairwise comparison but now, they had to select an intensity of relative importance (Table 6). According to Table 2. The time span between the first and the second pairwise was around 12 min; however there were few differences.

Criteria weights were determined on the basis of assigned points and comparison matrix. Differences between preferences are shown in the following Table (Table 7).

Table 7 Criteria weight comparison

	Fuller	Scorring	Saaty
Duration of the project	*0.1515*	0.1302	0.1672
Community	0.0379	0.0681	0.0748
Commercial support	0.1136	0.1050	0.1237
Documentation	*0.1515*	*0.1316*	*0.1730*
Trial/demo version	0.0758	0.0855	0.0782
HW requirements—RAM	0.0379	0.0402	0.0324
HW requirements—CPU	0.0379	0.0483	0.0268
Supported OS	0.0682	0.0680	0.0584
Integration	0.0697	0.0740	0.0660
Configuration	0.0455	0.0770	0.0746
User interface	0.1136	0.1050	0.0651
Language	0.0833	0.0671	0.0659

Table 8 Overall ranking of criteria weights

Fuller	Saaty	Scorring
C1	C4	C4
C4	C1	C1
C3	C3	C3
C11	C5	C11
C12	C2	C5
C5	C10	C10
C9	C9	C9
C8	C12	C2
C10	C11	C8
C2	C8	C12
C6	C6	C7
C7	C7	C6

The most significant criterion is C4—Documentation, which has the best result for all utilized methods. On the other side, the less important criteria are related with HW requirements. These criteria have the lowest weight for all methods. The overall ranking of final weights is described in Table 8.

Final results (Table 9). Show that Saaty's method has the most similar result as an evaluation based on users' experiences [13]. However, if standard deviation will be included into the overall evaluation, then all results go down well.

Table 9 Overall ranking of criteria weights

Method	Zenos	Nagios
	(%)	(%)
Scoring	79.80	83.05
Saaty	*81.85*	*84.36*
Fuller	78.91	81.82
Users	*82.8 ± 11*	*95.6 ± 3.6*

5   Conclusion

Due to the widespread of information and communication technologies, which today affects absolutely all human activity, is the use of IT management absolute necessity. ITIL® framework has deal whit this issue with more than 20 years of experience. It gathers the best experience in IT management and provides advice and tips on how companies can improve overall IT management efficiency.

The main objective of this research was to design a procedure for evaluating Free and Open Source ITIL tools. Licenses for commercial products are often going up to the order of hundreds of thousands of Czech crowns. And even so there is no guarantee that the product purchased for individual company is the right solution. Another option is to choose from Free or Open Source solutions. However, they are on the rise and each year comes a large amount of new projects. Not all of them have high quality and have a future. Another fact is the absence of a database or a web portal, which would be devoted to the issue. Based on these fact was created project about the evaluation of Free and Open Source ITIL tools. This project and all results may help with proposal of methodology for evaluation of ITIL tools which is main objective of author's doctoral thesis.

Acknowledgements This work was supported by the Ministry of Education, Youth and Sports of the Czech Republic within the National Sustainability Programme project No. LO1303 (MSMT-7778/2014) and also by the European Regional Development Fund under the project CEBIA-Tech No. CZ.1.05/2.1.00/03.0089 and also by the Internal Grant Agency of Tomas Bata University under the project No. IGA/CebiaTech/2016/006.

References

1. Axelos: ITIL Continual Service Improvement, 2nd edn, TSO, London, xi, 246 pp. ISBN 978-0-11-331308-2. http://www.best-management-practice.com. (2011)
2. Axelos: ITIL Service Design, 2nd edn. TSO, London, xi, 442 pp. ISBN 978-0-11-331305-1. http://www.best-management-practice.com (2011)
3. Axelos: ITIL Service Operation, 2nd edn, TSO, London 2011, xi, 370 p. ISBN 978-0-11-331307-5. http://www.best-management-practice.com (2011)
4. Axelos: ITIL Service Transition, 2nd edn, TSO, London, xii, 347 pp. ISBN 978-0-11-331306-8. http://www.best-management-practice.com (2011)
5. Axelos: ITIL Service Strategy, Stationery Office, London, xii, 264 pp. ISBN 978-011-3310-456. http://www.best-management-practice.com (2011)

6. Bucksteeg, M.: ITIL 2011. 1. Computer Press, Brno, 216 pp. ISBN 978-80-251-3732-1. (2012)
7. Ho, W., Xu, X., Dey, P.K.: Multi-criteria decision making approaches for supplier evaluation and selection: a literature review. Eur. J. Oper. Res. 16–24 (2010)
8. Kralik, L., Senkerik, R., Jasek, R.: Proposal of evaluation criteria for free and open source tools for modelling and support of it service management according to ITIL. In: Proceedings —29th European Conference on Modelling and Simulation, ECMS 2015, pp. 537–542 (2015)
9. Cerny, M., Gluckaufova, D.: Multicriterial evaluation in practice. Statni nakladatelstvi technicke literatury, Praha (1982)
10. Fotr, J., Svecova, L.: Managerial Decisions: Processes, Methods and Tools. Ekopress, Prague (2010). ISBN 978-80-86929-59-0
11. Krupka, J., Kasparova, M., Machova, R.: Decision Processes. University of Pardubice, Pardubice (2012). ISBN 978-80-7395-478-9
12. Kralik, L., Senkerik, R., Nozicka, J.: Proposal of categories and availability of ITIL® tools. Int. J. Circuits, Syst. Signal Process. **9**, 222–226 (2015)
13. Kralik, L., Senkerik, R., Jasek, R.: Comparison of MCDM methods with users' evaluation. In: Sistemas y Tecnologías de Información: 11th Iberian Conference on Information Systems and Technologies (CISTI 2016). Gran Canaria, Španělsko: AISTI, 2016, s, pp. 491–495. DOI:10.1109/CISTI.2016.7521387. ISBN 978-989984346-2. ISSN 21660727

E-consultation as a Tool for Participation in Teachers' Unions

Carlos Quental and Luis Borges Gouveia

Abstract In the 2000s, the Internet became the preferred mean for the citizens to communicate. The YouTube, Twitter, Facebook, LinkedIn, i.e., the social networks in general appeared together with the Web 2.0, which allows an extraordinary interaction between citizens and the democratic institutions. The trade unions constantly fight governments' decisions, especially in periods of crisis like the one that the world, Europe and, in particular, Portugal are facing. In this regard, the use of e-participation platforms is expected to strengthen the relationship between trade unions and the education community. This paper reports the research about the planning and driving of a series of experiments of online public consultation, launched by teachers' trade unions. These experiments are compared with those of other countries, such as Australia, United Kingdom and United States of America. A quantitative analysis of the results regarding hits, subscriptions, and response rates is presented, and it is compared with the 90-9-1 rule, the ASCU model and data from government agencies. The experiments performed used the Liberopinion, an online platform that supports bidirectional asynchronous communication. A better understanding of the benefits of these collaborative environments is expected by promoting quality of interaction between actors.

Keywords E-participation · Public participation · Trade unions · Teachers unions · Unionism 2.0 · Digital mediation · Liberopinion

C. Quental (✉)
Instituto Politécnico de Viseu, Escola Superior de Tecnologia e Gestão, Viseu, Portugal
e-mail: quental@estgv.ipv.pt

L.B. Gouveia
Faculdade de Ciências e Tecnologia, Universidade Fernando Pessoa, Porto, Portugal
e-mail: lmbg@ufp.pt

© Springer International Publishing AG 2018
Á. Rocha and L.P. Reis (eds.), *Developments and Advances in Intelligent Systems and Applications*, Studies in Computational Intelligence 718,
DOI 10.1007/978-3-319-58965-7_11

1 Introduction

Nowadays, politicians begin to worry about the lack of interest, trust and participation of citizens in democratic politics. Consequently, governments and political institutions around the world are increasingly using the Internet in an attempt to revitalize democracy through online public consultation and citizen participation [1, 2].

The potential of online political communication was demonstrated in 2007 in the Australian elections, called "google election" by Gibson & Ward and, in 2008, by the Obama campaign, which brought great visibility to the use of electronic means [3]. Yet, it is relevant to note that most media used by politicians have few interactive features [4]. Goot states that, in the Australian elections, citizens participated online, on their own, in blogs and websites of activist groups like getup (getup.org. au), election tracker (electiontracker.net.au), you decide (youdecide2007.org), among others. Rainie and Smith claim that 46% of Americans used the Internet, during the Obama campaign, to access news about the campaign, share views and mobilize other citizens [5].

Saebo and Skiftenes identified as significant activities of e-participation, the electronic voting (e-voting), online decision-making, electronic activism (e-activism), electronic consultation (E-consultation), online election campaigns (e-campaign) and electronic petitions (e-petitions) [6, 7].

Despite the e-participation being especially highlighted within the political context, its application has also been studied in other organizations. The current paper presents a reflection on this new paradigm of civic participation within teachers' unions, for which no related studies were found. The trade union studied here is the largest teachers union in Portugal, the National Federation of Teachers (FENPROF), which is composed of the Teachers Union of the North, Centre, Greater Lisbon, South, Region of Madeira, Region of Azores and the Foreigner. From these, the main focus is given to the Teachers Union of the Centre Region (SPRC). The participation results and the tools employed are analyzed.

2 E-participation

The use of electronic media to support the government, and the central and local public administration, activities, allows the state to develop new forms of relationships with the citizens. Among these, it is relevant to highlight the direct and mediated digital interaction between each citizen, or group of citizens, and the different public bodies, which results in a new proposal for public participation called e-participation.

Several authors define e-participation, in general terms, as citizen participation in public service processes at different stages of the production chain (planning, decision making, implementation, evaluation), which according to them represents the slight difference between e-participation and e-Gov [8, 9]. Yet, we believe that the scope of e-participation is much broader and that it includes the participation of citizens in any public service.

Due to the crisis in the representative democracy, clearly visible in the declining participation in the elections [10], there is a considerable increase in the use of Internet tools for public consultation and citizen engagement.

According to [11], civic participation is the redistribution of power from the authority to the citizens. From the authors' point of view, there are eight types of public participation, which can be described as an 8-step ladder. These steps, all of which have been thoroughly discussed in the literature [12–14], range from manipulation to social control, and include intermediate steps such as information, consultation, delegation of power and partnership. Arnstein [11] argues that the closer the citizens are to the top of the ladder, the more capable they are of controlling their participation in the participatory procedures and of demanding intelligibility of the implemented procedures to those who participate.

The creation of public spaces that enable the civil society to express, conflict and negotiate, puts it in the middle of a process where the public actions cease to be the responsibility of the State, but are to be created and developed by actors instead, whose central objective is the promotion of democracy. As a result, citizens are to be allowed to democratically exercise their power through partnerships of public authority [15, 16]. This is the case of the current study.

2.1 E-consultation

Online consultations consist in using the Internet to inquire a group of citizens about one or more topics, thus allowing the sharing of information between participants through platforms on which they can make contributions, inform and influence policy and decision-making.

The e-consultations should take into account important aspects such as ensuring that all citizens can express their opinions about issues and policies that affect them; considering that the participants' time is valuable; keeping the portal or platform easy to use; showing the contributions; and allowing anonymity for those who wish, just to name a few.

The e-consultations have typically a well-defined period. The topics for discussions are pre-defined by the promoters and they may be moderated. Unfortunately, existing studies show that few citizens participate. Ferro and Molinari [17] concluded that only the activists (3–5% of the population) participate, based on the ASCU (Activists, Socializers, Connected, Unplugged) model. On the other hand, Cruickshank et al. [18] argue that 1% participation in any initiative of e-participation is considered a success. Considering a six-step ladder of participation (Creators, Critics, Collectors, Joiners, Spectators, and Inactive), a Forrester Research estimated that in the United States only 13% are Creators (people who publish in blogs, have web pages, upload videos), 33% are Spectators (only consume information, such as reading blogs and news, watching videos) and 52% are Inactive (do not participate in any online activity) [19].

Comparatively, 90% of Internet users are "lurkers" [20] and 9% contribute a little [21]. In an e-petition submitted to the Portuguese parliament about the payment of motorway tolls, the authors obtained an effective participation of 2.6% in a population of 593.084 inhabitants with Internet access [22]. In turn, 7% of the UK population participated in the "Downing Street-House of Commons Government e-petitions" initiative, and Rainie and Smith [5] claim that, during the Obama campaign, 46% of Americans used the Internet to access news about the campaign, share points of view and mobilize other citizens. Sebastião [23] state that the impact of online petitions in the political system, measured by the number of signatures, is low—only 6.8% resulted in discussion at the Portuguese Parliament.

There are several types of e-consultations. The simplest involves questions and answers in discussion forums integrated within a website, in which citizens are invited to post their opinions, questions and concerns, and are able to receive feedback from the promoters, who may or may not be governmental authorities.

The online polls are the second type of E-consultations, which essentially allow to quickly measure a specific question. Examples of this type of e-consultation include the online poll performed to teachers about measures to be taken against the government, which is discussed later, and the "Your Voice in Europe" of the European Commission.[1]

Although e-petitions are identified as e-participation activities, in some cases, they may also be considered another form of E-consultation [24], particularly if the participants are invited to discuss and enrich them. The e-petitions allow the citizens to interact and influence the policy decisions of governments or other political bodies, such as Parliaments. The e-panels are more sophisticated versions of online consultations [24]. A group of citizens is invited to exchange points of view through online discussion forums, polls, chat rooms or votes around a common theme or policy initiative. This is the case of a public participation initiative with FENPROF's leader, and other elements, in the Liberopinion platform.[2]

The most common type of E-consultation is the editorial consultation, in which citizens and civil society representatives are invited to comment, usually in the form of online discussions based on targeted policy documents. Another increasingly common trend points towards E-consultation initiatives that combine many of the aforementioned elements, either in the form of a complete portal dedicated to a specific political campaign or of multilevel features targeting different audiences. Some examples are the Liberopinion,[3] Madrid participa[4] or Ask Bristol.[5]

---

[1]http://ec.europa.eu/yourvoice/consultations/index_en.htm.
[2]http://fenprof.liberopinion.com/pergA.php?id=1,id=9,10,11andid=12.
[3]http://libertrium.com/.
[4]http://www.madridparticipa.org.
[5]http://askbristoldebates.com/.

2.2 E-consultations Versus Traditional Consultation

An important feature of e-consultations is that they give the opportunity to provide feedback and influence the political process outside the electoral cycle, i.e., not just during election campaigns, every 3 or 4 years. The online consultation is much faster than traditional communication between government and citizens [25]. The need for a head of a high-level department or even a minister imposes delays that are countless with the online nature [2].

For government institutions, the Internet promotes efficiency and effectiveness by reducing transaction costs [26]. For ordinary citizens, who have their own lives, e-participation has the practical convenience of online communication, the immediacy of communication, 24 h access and the location flexibility, assuming that citizens can reflect and participate in their spare time. Note that conventional meetings require a physical presence in a particular location, which implies short to long travels. Unlike traditional media, Internet applications enable multilevel communication. In the case of e-consultations, the Internet provides a level of reciprocity and involvement that would otherwise be difficult and expensive for institutions in offline communication.

Public consultations are associated with deliberative democracy, transcend geographical barriers and can accommodate large targeted groups more efficiently. In other words, they are a concept of virtual Agora where ordinary citizens, politicians and experts who are typically profiled in rigid power structures, can get involved [27]. In this regard, the e-consultations have a comparative advantage over their offline versions, for which the logistics of documents in paper format is costly, time consuming and possibly suppressed, if budgetary concerns arise.

The UK government, for example, has been successful in launching the "Downing Street, House of Commons Government e-petitions"[6] initiative to encourage civic participation—over three million signatures were collected (about 7% of the UK population) [24]. By the standards of e-participation, such participation rate is commendable. Yet, Smith and Dalakiouridou [24] point out that the expectations raised about the initiative (that the petitions will influence the debates in the House of Commons) are misleading since the petitions have no constitutional basis. The same thing happens in Portugal, as discussed in [23].

The European Parliament, on the other hand, is better placed because the resulting proposals are brought to the attention of Parliament's committees and/or other EU institutions.[7] The European Parliament has an everlasting committee and a set of procedures specified in Rule 192 of the Rules of Procedure of the European Parliament. Yet, the effectiveness of this process requires further research.[8]

---

[6]http://epetitions.direct.gov.uk.

[7]http://www.europarl.europa.eu.

[8]See petitions at http://www.europarl.europa.eu/aboutparliament/pt/00533cec74/Petições.html and European citizenship initiative at http://ec.europa.eu/citizens-initiative/public/welcome?lg=pt.

The potential for E-consultation seems to be increasingly feasible from the promoters' participation perspective. If better participation can be achieved using low-cost means such as the Internet, then the cost-benefit purpose of large scale consultations can be achieved. The potential advantages of an E-consultation approach over traditional consultation methods include: extensive involvement, with easier contribution; better quality of expressed ideas, opinions and proposals; greater perception of democratic legitimacy and of equality (citizens and promoters); better sensitivity to time constraints and the participants' levels of interest; production of statistically significant quantitative results; and lower costs of promotion and guidance toward unrepresented voices. Furthermore, it is based on a collective learning model, instead of a set of beliefs and knowledge.

3   E-consultations on Trade Unions. Unionism 2.0

According to Pinnock [28], the trade unions only belatedly recognized the potential of ICT, and when they adopted them, they limited themselves to storing, processing and disseminating information through computers and the Internet. The major obstacle against their implementation lies essentially within the organizations [29], both because of the required skills and, mainly, because of the changes that they may induce.

The trade unions generally invest time and human resources in more traditional communication tools. Since most organizations adopt a one-way communication, they do not exploit the Internet's full potential, and Rego et al. [30] argue that it is difficult to know the impact of the trade union sites because there are not many online mechanisms to provide an effective user involvement. It is, therefore, essential to promote inclusion, participation and transparency. Note, nonetheless, that the presence of the union organization among workers is still crucial, regardless of the technological means used.

The Internet, by allowing new methodologies and forms of communication, will lead these organizations to consult their partners before making major decisions and to implement bidirectional communication tools, thus contributing to their qualitative transformation. It is the Unionism 2.0.

In a paper about the unions' presence in the Internet, Correia and Marques Alves [31] state that only 59.6% have a website, from which only a few have online registration (interestingly, only teachers' unions offer it) and publish information regarding their activity, such as annual reports or statutes. The SPRC has had a portal for about 15 years,[9] which provides not only information, but also the legislation to teachers, statutes, activity plans and annual reports. The SPRC has discussion mailing lists among union representatives for decision making on several subjects and distribution lists for sending information or clarifications to its

---

[9]http://www.sprc.pt.

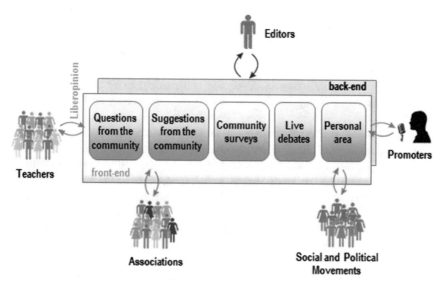

Fig. 1 Multidirectional communication between teachers and their trade union in the Liberopinion platform

members. During this research, the SPRC began a participation process that enabled its members to put questions about legal issues and make suggestions. Furthermore, they accepted the challenge of promoting their participation events using Liberopinion, an e-participation platform that enables effective participation in formal and informal deliberative processes via the Internet. The Liberopinion combines in a single, neutral and regulated, location, teachers and union representatives, and thus enables a multidirectional communication, as shown in Fig. 1.

4 Procedure

The e-participation projects need to have a well-organized structure [32]. In particular: (1) the actors to be included must be defined; (2) the issue to be addressed must be framed; (3) and a shared understanding that gathers consensus must be sought. The actors were chosen based on their representativeness, and the lack of studies in this field. Note that the FENPROF[10] represents about 70% of unionized teachers. The actors are composed of the trade union representatives on one hand, and of teachers, on the other hand. The participants' selection mechanism, chosen among the categories presented by Fung [33], was the diffuse public sphere. It is the most inclusive as it covers all people. On the other hand, still considering the same

---

[10]http://www.fenprof.pt.

categories, the participants can be framed in those that Fung calls mini-public and random selection, the best guarantee for population representativeness. Either way, all teachers can participate, even though the initial choice has fallen over the union members. The difference is that these are recorded automatically, and thus they have faster access to all content. Everyone can participate by making proposals, posting comments, voting on the questions posed by others, or following other proposals.

Given the low participation levels in political and community lives, it is necessary to plan ways of dissemination to raise awareness and encourage citizens to participate [1]. Several means are used to promote the platform and the debates: dissemination through the Trade Unions portal; dissemination via email to teachers, who are enrolled in the mailing lists, and members; and dissemination through the social networks. The information concerning the platform access statistics is stored in databases and logs for further processing using software such as Awstats Log Analyzer and Google Analytics tools.

This paper presents three events performed with teachers unions, such as petitions and public consultations. These events, described next in further detail, aimed, first, to listen to the teachers so that the unions could make decisions according to the teachers' wills, and, secondly, to influence the government decisions, particularly those of the ministry of education.

4.1   Liberopinion Usage Experiments: Education Manifest, Teachers Strike and Teachers Consultation

(1)   Education Manifest

The manifest "A School for a Portugal with future" aimed, according to an official, "to join a number of individuals around the fundamental idea that was crucial to change the political course that was being given to education by the minister of education". Teachers and researchers subscribed it. The trade union wanted to show the support given by people from the academic and political lives, and teachers in general, to the government using the slogan "As civically engaged citizens, people are today extremely concerned with the current course of the Portuguese education". For that purpose, the members appealed for the citizens and the structures of the Portuguese society, interested in changing the education policy, to actively participate and contribute to the adoption of measures that would open up prospects for a better future. People were invited not only to subscribe the manifest, but also to advertise it, enrich it, and discuss it in their workplaces and civic participation places.

The aforementioned petition ran, online, from March to June 2013, but was interrupted because "more urgent things appeared". The announcement was performed through the mailing lists and the union's portal.

(2) *Teachers strike*

In November 2012 and June 2013, the teachers went on strike. For the strike of June, they organized a rotation schedule among themselves so that (1) the review meetings would not take place and (2) the costs would be divided by all. During these strikes the access statistics, either to the SPRC's portal or to pages specifically dedicated to record relevant data, were accompanied. In this case, only teachers registered in the portal participated.

(3) *Teachers consultation*

After the strike of June 2013, the FENPROF, together with other ten trade unions, launched a survey, lasting 15 days, to hear the teachers' opinions about the measures to be taken next. The main goal, according to those responsible, was to "adapt the fighting directions according to the position expressed by the subscribers". The consultation was both online and offline.

5   Results

(1) *Education Manifest*

According to Pordata, only 62.1% of the Portuguese population had access to the Internet in 2013.[11] Considering that the total population[12] is 10.514.844 and that 8.398.245 are over 18, the population eligible for this study is of 5.122.929.

The results of the manifest "A school for a Portugal with future" are shown in Table 1. The manifest was visited 5471 times, from which 4782 were unique. The number of subscribers reached 1346, which represents 28% of the visitors. Compared to other studies, the participation was substantial, but it was low if compared to the Portuguese people aged over 18.

(2) *Teachers strike*

On the strike of November 2012, the platform registered 14.613 visits and 10.279 unique visits, as shown in Fig. 2. The pages related to the strike recorded 7.968 visits and 5.679 unique visits, which represents, respectively, 67.9% of the 11.740 registered users and 71.3% of the unique visits.

During the strike of June, the platform registered 18.714 visits and 12.991 unique visits, as shown in Table 2, while the pages related to the strike recorded 7.409 visits, from which 5.174 were unique.

---

[11]http://www.ine.pt/xportal/xmain?xpid = INE&xpgid = ine_indicadores&indOcorrCod = 00063 49&contexto = bd&selTab = tab2.

[12]http://www.pordata.pt/Portugal/Individuos+que+utilizam+computador+e+Internet+em+percenta-gem+do+total+de+individuos+por+sexo-1142.

Table 1 Results of the manifest

Time	Unique visitors	Visits	Pages/hits
March	1361	1543	3355/5105
April	552	706	1593/2182
May	151	193	348/468
June	2718	3029	6960/9863
Total	4782	5471	12256/17618
Subscribers total	1346		

Fig. 2 Access of registered users to the platform and strike related pages on the strike of November 2012

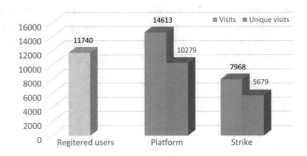

Table 2 Access to the platform and the strike

	Visits	Unique visits	Pages	Hits
Platform	18,714	12,991	177,029	580,186
Strike	7,409	5,174	11,682	17,870
% of visits	39.6%	39.8%		

(3) *Teachers consultation*

The questions shown in Table 3 composed the consultation performed. A total of 262 answers, distributed over the "Answers" column, was obtained online, which contrasts with the 8300 signatures obtained on paper, a preference already pointed out by Sebastião et al. [23]. In the open questions 2.1.3, 2.1.4, 2.2.4 and 2.2.5, several suggestions were performed, of which we highlight "concentrations in the Ministry of Education and Science (MEC)", "camp next to the MEC for a week, being each day ensured by a different region", "nightly vigils in front of the Parliament or MEC", "indefinite strike to prevent the beginning of the school year", and "thorough discussion about public education", just to name a few.

All answers were stored in a database. In addition to these, data regarding the school, the county and the district of each teacher was also stored. The visits were recorded in Google Analytics, depicted in Fig. 3, which monitored 8.326 unique visits and 11.645 visits, 11.472 of which were from Portugal, distributed by the regions of Coimbra (2564), Lisboa (1609), Porto (1207), Viseu (983), Aveiro (928), Setúbal (873), Leiria (709), Castelo Branco (551), Faro (464), Guarda (446), Braga

Table 3 Survey for listening the teachers and the number of online answers

Questions	Answers
1. Strike to reviews	
Continue the strike indefinitely	69
Continue the strike until the end of June	82
End the strike in the week of 17–21	106
2. Other forms of fighting	
2.1 To be performed in this school year	
2.1.1 Concentrations	167
2.1.2 Vigil	113
2.1.3 Several actions next to the MEC (Examples)	80
2.1.4 Others	53
2.2 To perform in September	
2.2.1 Concentrations	129
2.2.2 Manifestations	164
2.2.3 Vigil	164
2.2.4 Actions on schools (which?)	92
2.2.5 Others	39

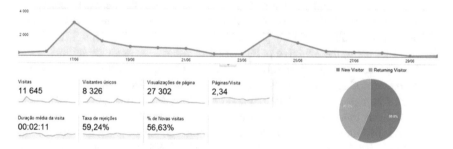

Fig. 3 Overview of visitors between 1 and 30 of June

(295), Santarém (247), Madeira (122), Viana do Castelo (107), Évora (88), Açores (80), Beja (79), Portalegre (68) and Bragança (52). The response rate was of 3.15%. Overall, mobile devices were used only by 7% of visitors.

6 Discussion

The results presented here show that the reference values have remained similar over time. The education manifest, which targeted the general population, was accessed less than 1%, while the remaining events, which focused on a more professional and educated group, with a high access level to technology,

Table 4 Summary table of the results of the initiatives

	Teachers' consultation	November Strike	June strike	Education manifest
Target population	163,175	11,740	51,833	5,122,929
Unionized teachers	81,587			
Unionized teachers FENPROF	51,833			
Visits	**7.1%**	**67.9%**	**14.29%**	**0.1%**
Visits versus total of teachers	7.1%	4.9%	4.5%	3.4%
Visits versus teachers unionized	14.3%	9.8%	9%	6.7%
Visits versus FENPROF	22.5%	15.4%	14.29%	10.6%
Signatures versus visits	2.3%			24.6%
Signatures versus total of teachers	0.16%			0.82%
Signatures versus teachers unionized	0.32%			1.65%
Signatures versus FENPROF	0.51%			2.6%

had participation percentages over 50%. In a report about the network society, the Centre for Communication (OBERCOM) states that the Internet use rate by citizens with a higher education is 96.9% [34], which may explain the high number of visits. On the other hand, the effective participation is rather low, as other authors report.

The teachers' consultation shows an interesting aspect. Although it was conducted by several associations or unions all together, which could allow a greater number of participations, the participation levels were not that substantial if compared with other events held only by FENPROF and/or SPRC. Possible causes for this result include: (1) FENPROF gathers more members; (2) the unions that integrate FENPROF have a more active action; (3) FENPROF has more visibility in the media (its leader appears many times on TV).

Does the data presented here support the 90-9-1 rule of Nielsen [21], which states that only 1% of users participate, regardless of the sample size? The effective participation in this study varies between 0.16% and 2.6%, which is not that much different from the aforementioned rule, as shown in Table 4.

As aspired, the teachers' consultation had an impact on the decision-making processes of the promoters, since some of the proposals came true. Furthermore, as stated by those responsible, "the goal was to adapt the fighting directions according to the position there expressed", which the FENPROF did. The impact on the government decisions was limited. As suggested by teachers, there is currently an ongoing petition, over the 20 regions of Portugal, to defend the public school.[13]

---

[13]http://www.escolapublica.net.

The limited use of mobile devices was consistent with the statistics presented in the literature, but further study is necessary since the increasing use of mobile devices could influence the citizens' participation in these type of initiatives. As stated by Ferro and Molinari [17], even if citizens have no Internet, they may have cell phones, and if they are connected, they may migrate to any of the ASCU model profiles.

7 Conclusion and Future Work

The analysis of the literature and of performed interviews indicate that no single solution exists for public consultation. A careful planning of methods and technologies to comply with the goals and needs of specific participants, such as teachers, and multi-platform approaches, such as the Liberopinion platform used by FENPROF/SPRC for online engagement, are likely to be the most effective [35].

Due to their foundation and organization, the trade unions will continue to put much emphasis on traditional forms of consultation, meetings with teachers, school debates, manifestations and concentrations in the streets. However, the authors defend that they are likely to lose their influence if they do not adopt new systems of public participation, like Liberopinion, because young people are increasingly focusing on online communication strategies.

The barriers to a greater online participation in policy making are cultural, organizational and non-technological [36]. In Portugal, the OBERCOM [34] states that 75% of respondents never participate in online polls and 72.3% never edit content. According also to OBERCOM [34], the Internet usage is a practice strictly related to each user's literacy level. These results are in agreement with Arnstein [11], i.e., that those involved are near the top of the ladder. Note that the results of the teachers' manifest are closer to the 90-90-1 rule of Nielsen, but, overall, the results obtained for the teachers' events are closer to those of OBERCOM [34] and Ferro and Molinari [17].

The citizens' participation is limited, as also highlighted by the low number of visits in other initiatives. For instance, the Global Parliament, a project created by large media groups, had 70.584 visits in 2012; from April to July, 2014, it had 117.518 visits, which represents an average percentage of less than 5% of visits. Nevertheless, the current study shows that: (1) the greater the focus, the greater the number of visits, which is consistent with other experiments carried out by the authors [37]—note that the number of visits represented 67.9% of the 11.740 SPRC members in the November teachers' strike, and 14.29% of the 51.833 FENPROF members in the strike of June, whereas it only represented 7.1% of the targeted population of 163.175, and 0.1% of a population of 5.122.929 in the teachers' consultation and education manifest, respectively; (2) the largest number of visits is from unionized members; (3) FENPROF has a substantial weight, which agrees with their size; (4) the effective participation and the percentage of lurkers are close to the reference values reported in the literature [18, 21, 38]; (5) there are more subscribers on paper than online, a preference already pointed by other authors; and (6) mobile devices are rarely used.

As future work, it is necessary to include the categories for the nonqualified or marginalized by the society in a new model of policy-making participation through the development of *m-participation* systems. Accordingly, the Liberopinion platform needs to be optimized to be used on mobile devices (smartphones and tablets). Considering that the report of OBERCOM [34] notes that 99% of people have TV, 88.5% own a mobile phone, and the Internet is the main route for information, there are still questions needing an answer: What strategies should be adopted to capture the attention of these users? What methods should be used for them to participate?

References

1. Dahlgren, P.: Media and Political Engagement: Citizens, Communication and Democracy. Cambridge University Press, New York (2009)
2. Macnamara, J.: The quadrivium of online public consultation: policy, culture, resources, technology. Aust. J. Polit. Sci. **45**(2), 1–16 (2010)
3. Gibson, R., Ward, S.: Introduction: e-Politics—the Australian experience. Aust. J. Polit. Sci. **43**(1), 1–11 (2008)
4. Goot, M.: Is the news on the internet different? leaders, frontbenchers and other candidates in the 2007 Australian election. Aust. J. Polit. Sci. **43**, 99–110 (2008)
5. Rainie, L., Smith, A.: The Internet and the 2008 Election. Pew Internet & American Life Project report, USA (2008)
6. Sæbø, Ø., Rose, J., Skiftenes Flak, L.: The shape of eParticipation: characterizing an emerging research area. Gov. Inf. Q. **25**(3), 400–428 (2008)
7. Medaglia, R.: eParticipation research: moving characterization forward (2006–2011). Gov. Inf. Q. **29**(3), 346–360 (2012)
8. Macintosh, A., Coleman, S., Schneeberger, A.: eParticipation: the research gaps. In: Proceedings of 1st international conference on electronic participation, ePart 2009, pp. 1–11 (2009)
9. Susha, I., Grönlund, Å.: eParticipation research: systematizing the field. Gov. Inf. Q. **29**(3), 373–382 (2012)
10. Quental, C., Gouveia, L.B.: Web platform for public e-participation management : a case study. Int. J. Civ. Engagem. Soc. Chang. (2014)
11. Arnstein, S.: A ladder of citizen participation. J. Am. Inst. Plann. **35**(4), 216–224 (1969)
12. Grönlund, Å.: ICT is not participation is not democracy—eParticipation development models revisited. In: Proceedings of 1st International Conference on Electronic Participation, ePart 2009, pp. 12–23 (2009)
13. Fung, A.: Varieties of participation in complex governance. Public Adm. Rev. **66**(1991), 66–75 (2006)
14. Christiano, T.: The Rule of the Many: Fundamentals Issues in Democratic Theory. Westview Press, Boulder, CO (1996)
15. Hansen, H., Reinau, K.: The citizens in e-participation. In: Electronic Government-5th International Conference, EGOV 2006, pp. 70–82 (2006)
16. Islam, S.:Towards a sustaineble e-participation implementation model. Eur. J. ePractice **5**(10) (2008)
17. Ferro, E., Molinari, F.: Making sense of gov 2.0 strategies: 'no citizens, no party'. JeDEM-eJournal of eDemocracy **2**(1), 56–68 (2010)

18. Cruickshank, P., Edelmann, N., Smith, C.: Signing a an e-petition as a transition from lurking to participation. In: Chappellet, M., Glassey, J., Janssen, O., Macintosh, M., Scholl, A., Tambouris, J., Wimmer, E. (eds.) Electronic Government and Electronic Participation, pp. 275–282. Trauner, Linz, Austria (2010)
19. Li, C.: Social Technographics: Mapping Participation in Activities Forms the Foundation of a Social Strategy. MA. Forrester Research, Cambridge (2007)
20. Lange, A., Mitchell, S., Stewart-Weeks, M., Vila, J.: The Connected Republic and the Power of Social Networks. Cisco Internet Business Solutions Group (2008)
21. Nielsen, J., Tognazzini, B.T.: Participation Inequality: Encouraging More Users to Contribute. http://www.nngroup.com/articles/participation-inequality/ (2014)
22. Quental, C., Gouveia, L.B.: Public participation employing digital means : will the politicians use new media with old strategies?, In: Actas de la 9ª Conferencia Ibérica de Sistemas y Tecnologías de Informacion, pp. 247–253 (2014)
23. Sebastião, S., Pacheco, A., Santos, M.: Cidadania digital e Participação Política: O Caso das Petições online e do Orçamento Participativo. Estud. em Comun. **11**, 29–48 (2012)
24. Smith, S., Dalakiouridou, E.: Contextualizing public (e)Participation in the governance of the European union. Eur. J. ePractice **7** (2009)
25. AGIMO: Consulting with Government—Online, Australian Government Information Management Office, Camberra, Australia (2008)
26. Tolbert, C., Mossberger, K.: The effects of E-government on trust and confidence in government. In: Public Administration Review, pp. 354–369 (2006)
27. Tomkova, J.: e-consultations: new tools for civic engagement or façades for political correctness? Eur. J. ePractice **7**, 45–55 (2009)
28. Pinnock, S.: Organizing virtual environments: national union deployment of the blog and new cyberstrategies. J. Labor Soc. **8**(4), 457–468 (2005)
29. Rego, R., Naumann, R., Alves, P.: Towards a typology of trade unions uses of the internet: preliminary data on the Portuguese case. In: 9th Congress of the International Industrial Relations Association (2010)
30. Rego, R., Alves, P., Silva, J., Naumann, R.: Os sítios na internet dos sindicatos portugueses: navegação à vista?, Sociol. Probl. e práticas 93–110 (2013)
31. Correia, M., Marques Alves, P.: ÁREA TEMÁTICA: Trabalho, Organizações e Profissões. In: Congresso Português de Sociologia (2012)
32. Staiou, E.-R., Gouscos, D.: Socializing e-governance: a parallel study of participatory e-governance and emerging social media. In: Reddick, C. (ed.) Integrated Series in Information Systems—Comparative e-Government, pp. 553–559 (2010)
33. Fung, A.: Varieties of participation in complex governance. Public Adm. Rev., 1991 (2006)
34. OBERCOM: Sociedade em rede: a internet em Portugal 2012. Publicações OBERCOM-Observatório da Comunicação, Lisboa, Portugal (2012)
35. Miller, L., Williamson, A.: Digital Dialogues: Third Phase Report Aug 2007–Aug 2008. Hansard Society, London (2008)
36. OCDE: Promise and Problems of E-democracy: Challenges of Online Citizen Engagement. OECD Publications Service, Paris (2003)
37. Quental, C.: A mediação digital como suporte para a participação no contexto dos sindicatos de professores—Proposta de um modelo com base na análise de experiências de e-participação. Universidade Fernando Pessoa (2015)
38. Preece, J., Shneiderman, B.: The reader-to-leader framework: motivating technology-mediated social participation. AIS Trans. Hum.—Comput. Interact. **1**(1), 13–32 (2009)

A Method for Quality Assurance for Business Process Modeling with BPMN

Waldeyr Mendes C. da Silva, Aletéia P.F. Araújo,
Maristela T. Holanda and Rafael T. de Sousa Júnior

Abstract Within the development of computer systems, the goal of business process mapping is to understand, improve, and organize the process activities to ensure efficient and correct implementation for releasing important firsthand inputs for requirements engineering. These inputs can be improved by process modeling. The quality of business process modeling can decisively influence the software quality. Internal quality analysis is done in-house, for the development process, while external quality analysis is done for the final product and satisfaction of the end consumer. Therefore, process modeling contributes to the production of consistent artifacts inorder to improve communications within the project. A way to assure quality comes from quality attributes verified with checklists in iterative cycles. Process mapping, then, contributes both to producing consistent artifacts and to improving communications within the project.

1 Introduction

Business processes are a collection of one or more activities performed following a predefined order to achieve an objective business goal, usually within the context of an organizational structure that defines functional roles or relationships [4]. Furthermore, a process can be regarded as work activities organized in a specific order in time and space, with clearly identified inputs and outputs [4]. In more general terms, a process can be defined as a set of activities that transforms resources (inputs) into results (outputs).

W.M.C. da Silva (✉)
Federal Institute of Goiás, Farmosa, Brazil
e-mail: waldeyr.mendes@ifg.edu.br

A.P.F. Araújo · M.T. Holanda
Department of Computer Science, University of Brasília, Brasília, Brazil

R.T. de Sousa Júnior
Department of Electrical Engineering, University of Brasília, Brasília, Brazil

© Springer International Publishing AG 2018
Á. Rocha and L.P. Reis (eds.), *Developments and Advances in Intelligent Systems and Applications*, Studies in Computational Intelligence 718,
DOI 10.1007/978-3-319-58965-7_12

169

Process mapping is widely used within organizations to understand organizational complexity and is an analytical and communication management tool that is used to establish the structure of existing processes and to conceive modifications for the improvement of these processes. Process mapping allows a better understanding of a process, mapping out exactly what it does, while Business Process Modeling (BPM) is focused on the optimization of business processes and can eliminate or simplify those processes needing changes [9]. The process modeling enables the evaluation and improvement of business processes [12]. Pidd [16] endorses the process modeling as a mechanism to discover which components are essential and improvements that will make a difference.

The task of mapping processes in a standardized and integrated way is particularly complex. In the context of modern software development, much emphasis has been placed on process mapping because of its importance and versatility.

Cost savings in product and service development, reduction in integration gaps between systems, and performance improvement in organizations are important results that may come from process mapping. This occurs because process mapping is an activity in which it is possible to identify business activities, providing a natural discovery of software requirements.

Giaglis [7] proposed a taxonomic classification for business process modeling and information systems modeling techniques. For him, the modeling is supported by methods which use techniques and many techniques are performed with tools. Recker [19] has expanded on this and proposed that the techniques of process modeling fall in two categories: mathematical paradigms like Petri Nets [14] and graphical modeling techniques such as Event-driven Process Chain (EPC) [20]. The Fig. 1 shows a BPM taxonomy that combines the proposals of Giaglis [7] and Recke [19].

Techniques based on graphics are more concerned with capturing and understanding the processes for the project scope of tasks and with discussing business requirements and process improvement with experts in this field [19]. In contrast, in contrast to mathematical approaches, that enable formal checks for correctness, graphical modeling techniques require human intervention to ensure the quality of the mapped process.

Fig. 1 Taxonomy of BPM

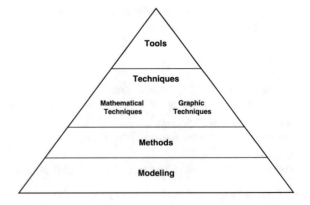

Freely from platforms, implementations and technologies, processes can be elucidated at a business level using a workflow-based notation. Many techniques are used for mapping processes in the same way, there are various graphical representations of the mapping process [19].

To graphically represent business processes, one of the most commonly used language has been Business Process Model and Notation (BPMN)[1]. BPMN is a standard for business process modeling with a graphical notation for specifying business processes in a Business Process Diagram (BPD). This graphical notation, based on a flowcharting technique, is very similar to activity diagrams from Unified Modeling Language (UML)[2]. BPMN aims to support business process management by providing an intuitive notation for business users and is also able to represent complex process semantics [15] and to facilitates the understanding of those involved, encouraging collaborations and business transactions in the organization [4].

The quality of a business process model highly influences the desired activity [12] as the quality of software is influenced by the quality of its documentation [17].

In large institutions, the processes modeling is a difficult challenge, since the knowledge of the processes to be modeled is distributed among employees who organize these processes in different ways. Usually the team responsible for business processes modeling consists of several people and this fact may yield different interpretations from the same workflow.

In the context of software development, the artifacts generated by process modeling with BPMN, are software documents, and can be submitted to quality controls. To deal with this fact, we have established a method for evaluating the quality of the artifacts generated by process mapping using BPMN.

2 Quality on Software Documentation

Software quality control assurance is closely related to verification and validation activities that are present throughout the software life cycle, including specific activities regarding documentation [13].

Documentation is an integral part of the software and its content is an important requisite to effectively develop the software. Unfortunately, in practice, the establishment of appropriate documentation is largely neglected [2].

The entire set of artifacts describing a software product, and all documents produced during the development process are considered software documentation. This includes documents of process modeling.

In recent years, about half of the research in process modeling consists of case studies or process design [18]. However, only a few studies have focused on methods to ensure the quality of the mapping process and its artifacts as [8, 12, 21, 23].

---

[1]http://www.bpmn.org

[2]http://www.uml.org

A quality measure is a quantitative scale, and a method that can be used in order to determine the value given to a characteristic of a software product [11]. To meet the needs of developers, maintainers, purchasers and end users, the requirements for the quality of the software product need criteria to evaluate the internal quality, external quality and quality in use [10] as defined below:

- *External Quality* is the totality of software product characteristics from an external point of view. It is the quality when the software is executed, which is typically measured and evaluated while it is being tested in a simulated environment with simulated data and using external metrics.
- *Quality in Use* is the quality idea of the software product from the user's point of view. This quality is measured when the software is used in a particular environment and specified context.
- *Internal Quality* regards all of the software product characteristics from an internal point of view. Internal quality is measured and evaluated against the internal quality requirements. The internal quality requirements, in turn, are used to specify the properties of the intermediate products. These can include static and dynamic models, other documents and source code [1].

Evaluations of software document artifacts are part of the approach to achieve internal quality. The idea of formal inspections was first presented by Fagan [6], who demonstrated the ability of this approach to produce significant improvements in software quality and developer productivity. In the original approach, error data were classified by type, and then occurrence frequency statistics were used to enhance the inspection process, in order to improve its efficiency.

Since then, there have been many changes, proposals and experiences described to check quality software documentation artifacts such as pair programming [3], review based on lists and user point of view [22], an approach based on views [2] and knowledge-based [5].

Specifically for BPMN, Khlif [12] established a correspondence between BPMN concepts and object oriented concepts by adapting object oriented metrics, obtaining new metrics which provided information about the complexity of business processes, cohesion between process tasks and coupling between processes themselves.

Regardless of the approach to quality assurance, desirable attributes are established, and their presence or absence is observed and/or measured. According Plösch [17], the *accuracy, precision, clarity, readability, structure* and *intelligibility* are considered the most important quality criteria for software documentation quality.

Accuracy is synonymous with correctness. Clarity signifies no ambiguity in the information contained in the documents. Consistency is the ability of the document to be free of actual or potential discrepancies, based on elements such as tables, figures and a glossary. Legibility deals with the presentation of the document, and is a consequence of the structure. Finally, intelligibility is an attribute that requires the presence of all the previous ones.

3 Process Modeling with Quality Assurance

Process modeling is also driven by processes which are divided into macro-processes, here called phases [4]. Process modeling comprises two main stages, with phases:

1. *Pre-modeling stage.*
2. *Process modeling stage.*

   i. SPMP—Startup Process Modeling Phase.
   ii. IPMP—Implementation of Process Modeling Phase.
   iii. VPMP—Validation Process Mapping Phase.

The *Pre-modeling stage* aims to define tasks that will be useful during the entire process modeling phase, even if the modeling involves different teams. The modeling itself occurs in the *Process modeling stage* and must be implemented complying with constraints on *Pre-modeling stage*.

Two teams are defined: the *modeling team* and the *quality team*. The quality team should be defined in such a way that its members are not involved in the modeling process. This requirement is to ensure that the mapped flow can be understood by an external team that necessarily has no prior knowledge of the modeled business process. However, there should be a senior reviewer, active in the modeling team and quality team, whose responsibility it is to semantically check the quality over of the whole process modeling.

The quality check is ongoing during and after the modeling. The semantic aspects are checked during the modeling. Quality assessment of the modeling occurs in the

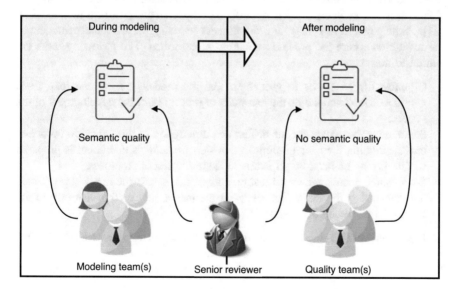

Fig. 2 Quality assessment roles

activity *Analysis and process improvement* in the IPMP macroprocess by the modeling team when they measure the semantic quality. The previous activities continue until the process is finished. The quality check during the modeling process, i.e., semantic checking, is expert-dependent and is performed by the senior reviewer. After completion of this process, it is forwarded to the *Quality Assessment Process*.

The quality assessment is carried out after the modeling occurs in the artifacts generated by the process modeling (process design and its description). The non semantic aspects are checked at this stage. This method for assessing the quality of the artifacts generated by the modeling process has been devised to solve this problem. The idea is to ensure that the flow is clear, objective, and within the set standard. The quality check is performed after the modeling process is carried out, with checklists by the quality team, which makes it less expert-dependent.

The completion of the phases occurs through the meetings' cycle. Records are kept with the main points discussed at each meeting. The phases generate results (defined as products). The progress of each phase depends on the quality of the products generated by the previous phase. An overview of a quality check with teams and a senior reviewer is shown on (Fig. 2).

4   Pre-process Modeling

This stage is critical to ensure standardization of the modeling processes and thus facilitate their integration. The artifacts generated by these phases are standardized documents to be used by all business process modeling teams. The central idea is to make the processes modeled by each team adopt the same set of BPMN symbols. This results in easy interpretation, regardless of the team that performs the modeling [21]. Participants in this stage are: the process modeling experts and managers of the institution where the process modeling is performed. The following tasks are central to this stage:

- Organize the processes in groups so that the modeling team can map them within a context and define the hierarchy of processes for the development of the system;
- Standardize the syntax for the design and description of the imposition process, thus facilitating their integration. In this step a standards document is prepared so that it may be used by all teams modeling business processes;
- Set a shared access, where all team members can access the mapped processes, thus increasing the interaction between the teams, and giving visibility to the work of other modeling teams.

The product of this stage is the standardization document to be used by all teams modeling business processes. The idea is that each team adopt the same set of BPMN symbols for the modeled processes, ensuring the standardization and easy interpretation, regardless of the team that has done the modeling [4].

Fig. 3 States of processes

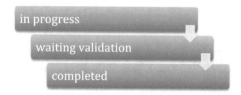

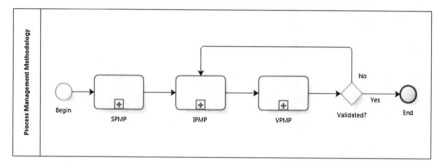

Fig. 4 Stages of *process modeling*

It is strongly recommended to have a network available environment to share access to documents, in which, over the Internet, the participants of the process modeling, properly authenticated, access the modeled processes, as well as information on the state of the current process, as shown in the Fig. 3. In this repository versioning services and wikis are tools that can be used.

5 Process Modeling

The Process modeling stage consists of three macro-processes (phases), as shown in Fig. 4. The macro-processes to be followed in this step are: SPMP—Startup Process Modeling Phase, IPMP—Implementation of Process Modeling Phase, VPMP —Validation Process Modeling Phase.

6 SPMP—Startup Process Modeling Phase

The Startup Process modeling Phase aims to introduce the participants of the modeling; set the agenda of meetings, defining the days of the week and times; present the method of work to be performed, including all activities internally defined in the pre-modeling phase; and define the macro processes and procedures which will be mapped in the IPMP. Thus, the products generated in this phase constitute the agenda of meetings and a project plan with schedule, scope, assumptions and constraints.

7 IPMP—Implementation of Process Modeling Phase

The Implementation of the Process Modeling Phase aims to map and model the business process modeling, describing each element. This stage is considered the core of process modeling, since the artifacts with the mapped processes are produced in this stage. It is essential to use an adequate tool for drawing procedures map. The products generated in this phase are the process design and its description. The process that represents the workflow of this stage is composed of the following activities:

- Process Modeling: in this activity, meetings are held for process modeling with the team and the technicians who have knowledge of the process to be modeled. Depending on the complexity of the process, this activity can take a week or even months;
- Analysis and Process Improvement: held weekly throughout the modeling process, the specialist team for modeling does flow analysis in order to improve the process.

The quality team should define checklists using artifacts quality criteria according to the standardization document generated in the pre-process modeling. The checklists are divided into two parts, one regarding the structure of the document and the other its contents. The checklist should enable the checking of artifacts as the form, content and questions need to ensure that the attributes of accuracy, precision, clarity, readability and structure are satisfied. A deadline is agreed between the modeling team and the quality team and then begins the evaluation of the quality of the artifacts generated by the process modeling.

Note that this evaluation of artifact quality is not performed on the semantic aspect, it only checks whether the generated artifacts are in accordance with the document standards, quality attributes and the objective. The assessment of the semantic aspect quality occurs during the modeling by the modeling team, this because the quality team is not involved in the modeling process, so they do not know the business process. The method for assessing the quality of the artifacts generated by the modeling process is illustrated by the highlighted elements in Fig. 5. It is composed of the following activities:

- Analyze document: the mapped process is analyzed by the reviewers, who must complete a checklist indicating any non-compliance found. The ideal in this stage is that at least two reviewers evaluate the same document independently, thereby producing at least two evaluations of the same document;
- Consolidate results: conducted by the authority responsible for the quality, which should consolidate the results into a single checklist. In case of conflict, a meeting to solve it should be held. If the artifacts are in accordance with the checklist the assessment of the artifacts quality is finished, otherwise the checklist is sent to the modeling team;
- Send checklist: the authority responsible for quality sends the consolidated checklist to the modeling team;
- Assessing corrections made: the team that mapped the process must assess the corrections made. If there is no compliance the above steps must be repeated.

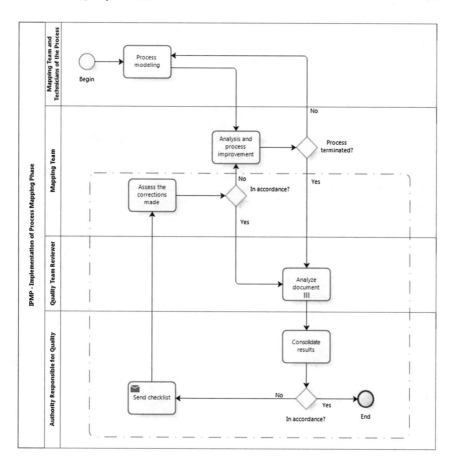

Fig. 5 IPMP processes

Thus, we note that for the implementation of this method the involvement of the following participants is required: reviewer, which can be formed by either a person or a team that will carry out a compliance analysis of the mapped process; the authority responsible for quality, which is the person responsible for consolidating the checklist sending it to the modeling team; the modeling team, which in this method will receive and analyze non-conformities found.

8 VPMP—Validation Process Mapping Phase

The VPMP aims to present the mapped processes to the managers in the field of business. This is the stage for acceptance of the product and it ends the life cycle of process modeling. At this phase, if the modeled process is not approved, or if there are any changes in the design, it returns to the previous step, as shown in Fig. 4

9   Case Study

The Personnel Management System of the Brazilian Federal Executive Power (SIGEPE) is the system to which this method was applied [21]. SIGEPE is a large computer system for personnel management of the Ministry of Planning, Budget and Management in Brazil. The related processes are predominantly in the area of human resources management. The numbers of SIGEPE are remarkable, with more than 60,000 function points distributed in 300 business processes. The development team of SIGEPE has about 500 people involved in process modeling [21]. SIGEPE is managed by the Ministry of Planning, Budget and Management (MPOG), the institution that establishes policies for than two million current and former federal employees.

One of the challenges we faced in this project was the large number of processes, the size of the team involved in modeling, the need for the process modeling teams to work simultaneously, in addition to different modeling requirements and development teams.

In this case study, the application of this method ensured uniformity in the quality of process modeling artifacts, maintaining a quality level greater than that stipulated by managers.

10   Summary

The presented method can be applied to the task of process mapping or process modeling. Semantics quality is expert-dependent, while the quality of the documentation is less expert-dependent. Since the quality criteria are established and checked by a quality team, independent of the modeling team, it is possible to ensure an established level of quality. We use checklists with quality attributes linked to each question to guarantee the criteria.

In the context of software projects, a consistent documentation is a key requirement for effective team communication. Artifacts with guaranteed quality levels collaborate for the internal quality and therefore the quality of the final product. The success of projects, especially those where communication is a critical factor, involves the application of a method that ensures documentation quality, including modeled processes.

In summary, this method uses well-defined stages that guarantee a level of quality for software artifacts—including modeled processes—having the quality attributes previously established as critera.

References

1. Azuma, M.: Applying iso/iec 9126-1 quality model to quality requirements engineering on critical software. In: Proceedings of the 3rd IEEE International Workshop on Requirements for High Assurance Systems (RHAS) (2004)
2. Bayer, J., Muthig, D.: A view-based approach for improving software documentation practices. In: 13th Annual IEEE International Symposium and Workshop on Engineering of Computer-Based Systems (ECBS'06), p. 10. IEEE (2006)
3. Bisant, David B., Lyle, James R.: A two-person inspection method to improve programming productivity. IEEE Trans. Softw. Eng. **15**(10), 1294 (1989)
4. de Sousa, R.T., de Deus, F., Aires, B., Ribeiro, G., Arajo, A.P.F., Holanda, M., Vidal, S.S.A. N., dos Santos, R.M.G., Cortes, F.P.: Business process modelling: a study case within the brazilian ministry of planning, budgeting and management. In: 2014 9th Iberian Conference on Information Systems and Technologies (CISTI), pp. 1–6. June (2014)
5. Ding, W., Liang, P., Tang, A., Van Vliet, H.: Knowledge-based approaches in software documentation: a systematic literature review. Inf. Softw. Technol. **56**(6), 545–567 (2014)
6. Fagan, M.E.: Design and code inspections to reduce errors in program development. In: Pioneers and their Contributions to Software Engineering, pp. 301–334. Springer (2001)
7. Giaglis, G.M.: A taxonomy of business process modeling and information systems modeling techniques. Int. J. Flex. Manuf. Syst. **13**(2), 209–228 (2001)
8. Hommes, B.J., van Reijswoud, V.: Assessing the quality of business process modelling techniques. In: Proceedings of the 33rd Annual Hawaii International Conference on System Sciences, 00(c), p. 10 (2000)
9. Hunt, V.D.: Process Mapping: how to Reengineer Your Business Processes. Wiley (1996)
10. ISO/IEC: ISO/IEC 14598International Standard c ISO/IEC Information Technology Software Product Evaluation—Part 1. ISO/IEC (1999)
11. ISO/IEC. ISO/IEC 9126: Software engineering—product quality. ISO/IEC (2001)
12. Khlif, W., Makni, L.: Quality metrics for business process modeling. In: Proceedings of the 9th WSEAS International Conference on Applied Computer Science, vol. 9(1), pp. 195–200 (2009)
13. Koscianski, A., dos Soares, M.S.: Qualidade de Software (2007)
14. Lohmann, N., Verbeek, E., Dijkman, R.: Petri net transformations for business processes–a survey. In: Transactions on Petri Nets and other Models of Concurrency II, pp. 46–63. Springer (2009)
15. Object Management Group (OMG): Business Process Model and Notation (BPMN) Version 2.0, vol. 50 (2011)
16. Pidd, M.: Modelagem Empresarial: Ferramentas Para Tomada de Decisao. Bookman (1996)
17. Plösch, R., Dautovic, A, Saft, M.: The value of software documentation quality. In: 2014 14th International Conference on Quality Software, pp. 333–342. IEEE (2014)
18. Recker, J., Mendling, J.: The state of the art of business process management research as published in the BPM Conference. Bus. Inf. Syst. Eng. **58**, 1–18 (2015)
19. Recker, J.C., Rosemann, M.M., Indulska, M., Green, P.: Business process modeling: a comparative analysis. J. Assoc. Inf. Syst. **10**(4), 333–363 (2009)
20. Scheer, A.-W., Thomas, O, Adam, O.: Process modeling using event-driven process chains. Process-Aware Inf. Syst. 119–146 (2005)
21. De Sousa, B.A., Freitas, H., Holanda, M., Mendes, W., Moraes, A., Silva, C., Vidal, S.S.A. N.: A methodology for quality assurance for business process modeling with BPMN. In: 11th Iberian Conference on Information Systems and Technologies (CISTI) (2016)
22. Thelin, T., Runeson, P., Wohlin, C.: An experimental comparison of usage-based and checklist-based reading. IEEE Trans. Soft. Eng. **29**(8), 687–704 (2003)
23. Trkman, P.: The critical success factors of business process management. Int. J. Inf. Manag. **30**(2), 125–134 (2010)

Performance Analysis on Voluntary Geographic Information Systems with Document-Based NoSQL Database

Daniel Cosme Mendonça Maia, Breno D.C. Camargos and Maristela Holanda

Abstract With the advent of Web 2.0 and mobile technology, including, smartphones equipped with GPS (Global Positioning System) receptors, there has been an increase in the number of individuals who create and share spatial data. Consequently, the ability to store a large quantity of data, in diverse formats is made possible by Geographic Information Systems (GIS) that use voluntary data. This study presents a performance analysis of data storage architecture of a Voluntary Geographic Information System (VGIS) that uses a document-based NoSQL database and a relational database for comparison. To carry out the performance analysis it was necessary to remodel the application database from a relational database to a non-relational model. Furthermore, insertion and reading tests were needed, and performed in local and clustered environments using a simulator that generates random data on a large scale. The test results have sought to analyze the performance and feasibility of using a document-based database from data storage architecture for VGIS.

1 Introduction

With the rise in the use of web services and mobile technologies, such as, smartphones equipped with GPS (Global Positioning System), there has been an increase in the number of individuals who generate and share spatial data. Creating geographic information, a function traditionally reserved for officials within restricted agencies, is now being carried out voluntarily by regular citizens who may have little

D.C.M. Maia (✉)
Federal Institute of North Minas Gerais (IFNMG), Pirapora, Brazil
e-mail: daniel.maia@ifnmg.edu.br

B.D.C. Camargos · M. Holanda
University of Brasilia (UnB), Brasília, Brazil
e-mail: breno.diogo@hotmail.com

M. Holanda
e-mail: mholanda@unb.br

© Springer International Publishing AG 2018
Á. Rocha and L.P. Reis (eds.), *Developments and Advances in Intelligent Systems and Applications*, Studies in Computational Intelligence 718,
DOI 10.1007/978-3-319-58965-7_13

or no formal qualifications in the field of geography. The GIS (Geographic Information Systems) that make use of data collected by voluntary users are called VGIS (Voluntary Geographic Information Systems).

A VGIS must have a storing system that supports the following characteristics: a large volume of data; many concurrent read and write operations; a large number of users; and heterogeneous data, both structured and unstructured, coming from a variety of sources. Faced with these challenges, it has become necessary to investigate how the data of the application can be stored, managed and shared efficiently in a digital environment [7, 14].

Although, the main data storage technologies in GIS use relational databases, the NoSQL databases have been increasingly utilized to handle large volumes of data. They also enable the architecture to be sufficiently flexible to prevent a potential increase in the number of application users [43, 44]. Thus, this Chapter presents 'Consulta Opinião' a Voluntary Information System with a data storage architecture that uses geographic data in voluntary geographic information systems using document-based NoSQL database.

This Chapter is organized as follows. In Sect. 2, Voluntary GIS (VGIS) and NoSQL databases are presented. Section 3 lists some related works; Sect. 4 presents the data storage architecture in VGIS using NoSQL Data. In Sect. 5, the details of the Case Study, are described, as well as, the application used, and the data modeling of the project carried out with the NoSQL Database. Section 6 presents the performance tests made in the application persistence layer. Finally, Sect. 7, presents the conclusions.

2   NoSQL Databases and Voluntary GIS

2.1   NoSQL Databases

NoSQL database systems emerged from the need to store large amounts of data, especially in web applications. Web applications permit a large number of concurrent users, which can compromise the performance of the DBMS (Database Management System). The NoSQL solutions are designed to meet the on-demand availability and scalability requirements of large-scale applications [32, 40].

In general, the NoSQL systems have the following characteristics [21, 33, 37]: they are non-relational; rely on distributed processing; provide high availability and high scalability; they are dynamic schema instead of static schema, which provide the flexibility; and have the ability to handle structured and unstructured data.

Currently, there are more than 225 NoSQL Databases cataloged, which were classified according to the data model used. Typically NoSQL databases are classified into four categories [13, 21]:

- Key-value stores: data is stored as a key-pair value. These systems are similar to dictionaries where the data are addressed by a single key. Values are isolated and independent from others, where the relationship is handled by the application logic. The simplicity of the key-value storage makes it ideal for recovery values in applications that have user profiles, session management or recovery product names such as 'shopping cart'. Some examples of key-value based databases are DynamoDB, Azure Table Storage, Riak, Redis and Voldemort [12];

- Column family databases: the column family database defines the structure of values as a predefined set of columns. The column family store can be considered as a database schema, where the super columns and column family structures determine the schema of the database. It is recommended to use the column-based database for distributed data storage, especially data under version control; large applications, data processing oriented lots; and predictive and exploratory analysis of statistical data. Some examples of column-based databases are Hbase, Cassandra, and Amazon SimpleDB Hypertable [32];

- Document-Oriented Storage: a document store uses the concept of a key-value store. The documents are collections of attributes and values, where an attribute can be multivalued. The keys inside documents must be unique. Each document contains an ID key, which is unique within a collection and identifies the document. The documents can be nested, i.e., a document can be referenced as value of a key in another document. A document has a free scheme, documents with different structures can be stored in the same collection. Document-based Databases can store structured data such as CSV (comma-separated values); semi-structured data as XML (eXtensible Markup Language) and HTML (HyperText Markup Language) files; and unstructured data, such as image files, audio, video, PDF (Portable Document Format) and text. Some DBMS of this type have the functionality to extract metadata binary files, which can be interesting for indexing and to provide a more efficient management of files (documents) stored [16]. The document-based databases are suitable for problems involving different environments and handle the storage of large volumes of data. In addition, it is recommended to store documents as denormalized conceptual representation of a database entity, and semi-structured data that requires the use of null values [35]. The use of document-based databases is also recommended for event logging (logs); content management and blog platform systems; Web analysis in real time; e-commerce applications that should have flexible storage schemes. Some examples of document-based databases are MongoDB and CouchDB [37];

- Graph databases: a graph database uses graphs to represent schema. Graph Database works with nodes and edges between nodes. Graph-based Databases have small records with complex interconnections. So their use is indicated when you need to represent relationships such as social networks, detection of fraud and applications that require high performance in queries with many joints. Some examples of graph-based databases are Neo4J, Infinite Graph, Sparksee, TITAN and InfoGrid [1, 29].

It is important to note that one should consider the adoption of a particular type of NoSQL database, according to the characteristics of the application and the data to be stored.

2.2   Voluntary GIS (VGIS)

Changes in technology are altering the traditional way geographic information is produced, which is primarily done by national mapping agencies that use costly means. The Web 2.0, is characterized by its capacity to allow users to create content and interact with content from other users in real time—georeferencing, geotagging, GPS, wide band connection, as well as other technologies, which has provoked the rise of voluntary geographic information [11, 17].

The VGIS makes use of tools for creating, gathering and disseminating geographic data derived voluntarily by individuals. All humans are capable of acting as intelligent sensors, using the most diverse equipment to obtain mediations [5, 9, 10].

The voluntary user is beginning to relocate and redistribute mapping agencies production of geographic knowledge to networks of actors in NGOs. Participants who produce are both producers and users, aka 'produsers' [4, 10].

The motivation to participate in VGIS has always been questioned, and one of the motives is to feel a sense of belonging within a community [24]. Along with this, in discussions about the motivation to use VGIS, the value of knowledge created also arises. Clearly, there are two areas in this line of questioning. On the one hand, some researchers acknowledge the benefits of the ability to map at the local level by using people who live in the particular environment studied [18, 39]. However, others are pessimistic about the capacity of individuals to disseminate information without being required to confirm the validity of the information, arguing that this will result in a chaotic mix of useless information, possibly compromising truth [23, 24].

The data created from a VGIS has its questionable credibility [15], as non-professional users and malicious people can post incorrect or false information, the systems must have verification mechanisms and validation of contributions made by volunteer users.

There are many discussions about the quality of volunteers geographic data, especially with regard to their accuracy (accuracy) and validity (credibility) [15]. The evaluation of quality of geographic information follows a set of protocols and criteria. The ISO (International Organization for Standardization) describes standards for data quality (ISO 19157) and special metadata (ISO 19115) [19, 30]. The parameters that define the quality of the data are completeness, logical consistency (integrity), positional accuracy, temporal accuracy, and thematic accuracy (fitness for use) [34]. Metadata about data quality are important to VGIS, given that some scientists dismiss voluntary data because they lack such information [2, 30].

Even faced with the risks, the strategy of using voluntary labor in GIS is valid because obtaining and maintaining spatial data is labor intensive and expensive [28]. Moreover, it is possible to obtain a particular view of the feelings of citizens about certain political and social issues in this way.

The VGIS is related to different interests, according to the target audience. From the industry's perspective, interest in VGIS is related to the development of appropriate tools for managing spatial data, with the possibility of proposing ways to verify and validate the volunteers' geographic data. The government's interest may be the adoption of public policies that seek to solve local problems, such as the mapping of diseases, crime patterns, noise levels; improving sanitation in disenfranchised areas; providing aid during disasters; among other uses for the benefit of urban planning. From an academic perspective, there are research fields that are concerned with the quality of data, visualization of geographic information, the best ways to handle large volumes of heterogeneous data, and how the information can be stored, managed, searched and shared efficiently [14, 34].

A prime example of VGIS is the OpenStreetMap, which is a global mapping project that aims to provide a set of free maps to use, and edit, and which is licensed under new copyright schemes. A main motivation for the OpenStreetMap project is to provide free access to geographic information [20].

3 Related Works

The adoption of the NoSQL Database for storing geographic data has been studied in the literature. Some of these related works are presented below.

In Zhang et al. [44], a storage approach of spatial big data is shown through the NoSQL database based on MongoDB document. MongoDB is a database management system that provides high performance and extensible data storage methods for applications with large amounts of data. In addition, it shows good performance in data query. The article presents the approach used to store, query and update spatial data in MongoDB using ESRI shapefile files and a script in Python. The aim is to read and analyze ESRI Shapefile (.shp, .shx, and .dbf) to generate a document that will be stored in MongoDB, containing the type of geometry, attributes and coordinates.

In Baptista et al. [3], relational databases are considered unsuitable for handling large amounts of data, and rather, the document-based NoSQL databases CouchDB and MongoDB are appropriate for providing support for the storage of spatial data. The article proposes an architecture to solve interoperability problems in data storage between relational database systems and NoSQL, through the implementation of OGC (Open Geospatial Consortium) services such as WMS (Web Map Service) and WFS (Web Feature Service). It is also cites the work of Miller et al. [31] using CouchDB to store spatial data and mobile interface for recovery data.

Vitolo et al. [43] suggested the use of a multidimensional database, such as, Rasdaman for environmental and climate scientific applications that can use multidimensional raster data. The use of variations of XML developed to deal with environmental and geographic data, such as WaterML and GML (Geography Markup Language) is also recommended. Furthermore, the article affirms that the construction of environment virtual observatories requires the use of various tools for data acquisition and analysis, and presented the EVOp project, which is a set of web applications that use environmental information from models and the local community tools.

Jardak et al. [22] proposes the use of the column-based NoSQL database Hbase and of the framework Hadoop for implementing processing structures for big spatial data; however, one must pay attention to some details in the architecture implementation of the project to benefit from the high degree of parallelism enabled by Hadoop. Ma et al. [26] also recommended the use of Hbase for implementing GIS with vector data or image data, such as remote sensing systems.

In Lizardo et al. [25], the column-based NoSQL database Cassandra is used to store geospatial data in a spatial database prototype, called GeoNoSQL. However, until the time of writing this work, Cassandra does not have mechanisms or native extensions to work with geospatial data and therefore is not able to index multidimensional data. The study uses the information recuperation library, Apache Lucene, with a geospatial extension, Lucene Spatial, to verify the spatial index of Cassandra. The solution adopted for the Cassandra data store can be applied in the construction of GISs that handle large amounts of data and require high scalability and performance.

In Pourabbas [36], the CouchDB, MongoDB, Bigtable and Neo4j Databases are referenced as examples of NoSQL systems that support geospatial data. In this case, a discussion about desirable resources for the spatial database systems was elaborated and a comparison between the NoSQL Databases that deal with geographical data. The desirable resources cited were, the following: support for the different systems of space references (SRID); the possibility of indexing spatial data; types of topological, metric and set data, among others.

4   Data Storing Architecture

The Voluntary Geographic Information Systems have characteristics that require appropriate strategies for efficient data storage. It is necessary to deal with issues involving the storage of a large amount of data and simultaneous access to multiple users, which can cause a lot of read and write operations.

Another problem inherent to VGIS is the diversity of formats that can be stored in applications. The database system adopted by VGIS should allow flexibility in the types of allowed data storage, especially considering that spatial data can be stored and image files, audio and video.

Faced with these challenges, it is possible to suggest that VGIS require a robust data storage architecture that adopts alternative technologies, such as banks NoSQL data for efficient data storage. Currently, spatial data storage NoSQL databases have been explored by researchers with the aim of creating a distributed architecture in the database, and supports the storage and processing of a large amounts of information.

The present study shows an alternative, which is an abstract architecture for data storage, in diverse formats, in VGIS, as shown in Fig. 1. This architecture has the function of storing data from an application in a DBMS.

This abstract architecture presents an organization in three layers: Presentation, Business and Data Persistence. The layers of Presentation and Business compose the

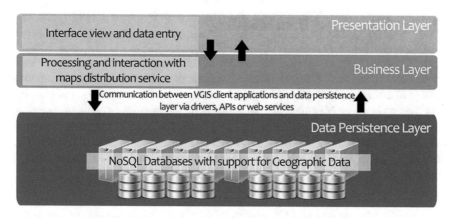

Fig. 1 Abstract Architecture for storing data in VGIS using a NoSQL Database

structure of client application, i.e., the VGIS that will be integrated with the NoSQL Database systems of the present in the Data Persistence layer. The following is a description of the items of the architecture:

- Presentation layer: in this layer the application should make interfaces available that are supported by mobile and desktop devices.
- Business layer: this layer completes functions of processing and communication with the NoSQL Database System. The communication of this layer with the system of NoSQL Database is carried out through the drivers specific to the programming language used in the implementation of the VGIS.
- Persistence Data layer: this layer is responsible for storing the data and metadata of the application in a DBMS. The proposal of this layer is to make possible the storing of data alpha numbers (conventional), geographic, and files in the format of images, audio and video. To support the characteristics of the VGIS, this architecture proposes the use of NoSQL Database giving emphasis to the scalability and heterogeneity requirements. The requirements of scalability aim to confront challenges, such as, frequent concurrent read and write operation in the database, and access to a great quantity of data; since the requirement for heterogeneity looks for the application to have support for storing in diverse data formats.

This data storage architecture for VGIS adopts document-based NoSQL databases, because this type of database provides greater robustness in the spatial data storage, natively supporting the standard for exchange of geospatial data GeoJSON; and also supports storage of a wide variety of data, which may load any type of document without needing to know the file structure in advance. The document-based databases also provide consultation on all data stored in the database, and an interesting feature for any application.

Adopting the distributed database technique, its possible to benefit from the storage and management of distributed data architecture. Thus, the NoSQL database systems can provide greater storage capacity, increased competition in reading and writing, as well as better availability and fault tolerance [26].

5 Application and Data Modeling

The 'Consulta Opinião' project [6] is a VGIS that was used to validate the storage architecture presented. This project aims to obtain data about the opinion of the users and public servants, and presents it to the managers of these services. For this, the application has a mobile interface (an Android application for smartphones) geared toward the use of public service clients, who completed the evaluation of the establishments registered in the system; and a web interface (developed in PHP), aimed at managing public services, which is expressed in a visual and dynamic way, through the map of the evaluations completed by voluntary users.

Systematically, the application works as follows:

- Public service users should be geographically close to the establishment that is being evaluated. In this case, the evaluator's smartphone GPS coordinates will be compared to the geographic coordinates of the public institution to be evaluated, which is stored in the database (Fig. 2).
- The evaluating user, through his or her smartphone, fills in the survey evaluation of the public establishment, and then sends this data to a central database. In this particular case it is possible to fill out up to 5 questions for each public institution. The questions can vary in content depending on the type of property to be evaluated (school or hospital). Thus, many aspects can be evaluated.
- Through a web interface, the public manager can then communicate with the central database and use the data from evaluations of voluntary users to direct management decisions (Fig. 2). The average score for each establishment may vary from 0 to 5 points.

Fig. 2 Mobile and Web interface of the application 'Consulta Opinião'

Originally the 'Consulta Opinião' project used the relational DBMS PostgreSQL in the central database, but was adapted for communication with the NoSQL Database MongoDB. The MongoDB was used in this application because it provides some characteristics considered essential for the VGIS; it has native support for the storing of geographic data; it carries out consulting functions and spatial analysis; is very consistent; and has simple data replication architecture, allowing the system to adapt to working with many users and a large quantity of data. Another positive point for adopting the MongoDB technology is its GridFS specification, which stores files in small pieces that afterward can be carried according to necessity, which avoids having to carry all of the file memory. The GridFS specification also is useful for storing files that exceed 16 megabytes of data [8, 36].

'Consulta Opinião', developed in a modular form, basically, required only the modification a layer of the applications persistence data for the MongoDB database.

Figure 2 shows the mobile interface used to select which public establishment will be evaluated by the application user; and the web interface used by the public administrator to view the information posted on the application by the volunteers.

Due to the change of the database from relational to a non-relational database, there was, consequently, the need to redo the modeling of data aimed at providing the efficient storage of the information in the system in a NoSQL Database. The model of data in non-relational projects can use the denormalization strategy, which incorporates data from tables of relational models in one sole document and generates a data redundancy control. The normalization strategy is used to prevent a query of data on different machines, which can generate traffic information in the network and information inconsistency. Thus, it creates a more complex document, which has all the information necessary to meet a specific request of the user, but that does not compromise system performance.

Figure 3 shows the relational model of the central database of the 'Consulta Opinião' project in its original architecture. The model consists of five tables: User,

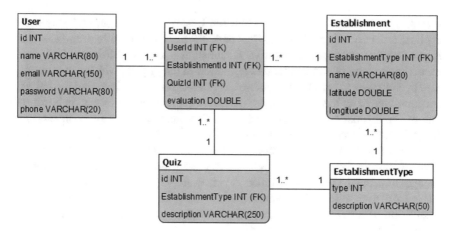

Fig. 3 Relational model of the 'ConsultaOpinião' project, according to the original persistence data layer

Fig. 4 Diagram of the
non-relational model of the
'Consulta Opinião' project
(embedded data model)

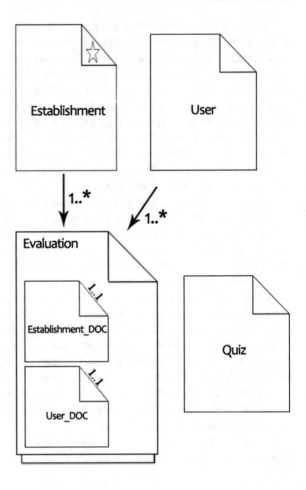

Establishment, EstablishmentType, Quiz and Evaluation. The User table stores the
application's user information. The Quiz table stores the questions registered for each
type of system establishment. Tables Establishment and EstablishmentType store
information about public institutions registered in the system, which can be evaluated by volunteer users. On the Establishment table it is possible to find the latitude
and longitude fields to store the geographic coordinates of the registered public institution. The Evaluation table stores the notes of the answers given by volunteer users
for each question on the reviewed establishments.

Figures 4 and 5 shows two versions of the non-relational model for the database
of the 'Consulta Opinião' project adapted to the document-based NoSQL database
MongoDB using the notation specified in Vera et al. [42]. In both versions, the model
has to store four collections (compared to the tables in the relational model): User,
Establishment, Quiz and Evaluation. Similarly the User table, the User collection
stores the application's user information.

Fig. 5 Diagram of the non-relational model of the 'Consulta Opinião' project (referenced data model)

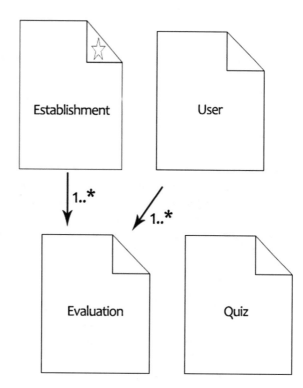

The Quiz collection also has the same function as Quiz table (relational model) and stores the questions registered for each type of establishment. The Establishment collection stores the information of the public institutions that can be assessed. In this collection geographical data is stored, specifically the geometric object point referencing the geographic coordinates (longitude and latitude) of the registered public institution. Unlike a version of a non-relational model to another, it is in the form of storing the data of the assessment made by the voluntary user. In the first version (Fig. 4), the evaluation collection incorporates data of Establishment and User collections, beyond the average grade assessment for each establishment. This strategy seeks to make a controlled data redundancy, in order to improve query performance and to facilitate the distribution of data on a possible deployment in a clustered environment. In the second version (Fig. 5), the evaluation collection only refers to collections Establishment and User through of these collections identification code, and records the average score of evaluation of establishment.

The notation presented in [42], shows the documents that manipulate geographic data using a star on their behalf. In this case, as can be seen in Fig. 4, the 'establishment' document contains this statement to represent that it stores the geographical location (point) of the registered public institutions in the application database.

For storing geographical data in a document-based NoSQL MongoDB database used a GeoJSON notation. This notation is an open format for exchanging geospatial data, based on JSON (JavaScript Object Notation), which may represent several geometries, such as point, line and polygon.

6   Application Tests

The application tests, together with a data storage architecture, were performed in a simulated environment. In this case, some users and establishments (schools and hospitals) were registered. Afterwards, to carry out the evaluations of the establishments registered, an evaluation simulator was developed, which generates a random grade for each measured requirement of the establishment and includes the data in the central DBMS. The simulator operates as follows:

1. A user registered in the database is randomly selected to conduct the evaluation of the establishment.
2. A registered public establishment is also randomly selected for reviews.
3. The simulator generates a random score for each question to be evaluated in establishing and then calculates a final average, from the notes generated for evaluation of the institution.
4. Store the data in the database.

The test scripts were developed in PHP language, having a runtime registration mechanism (in seconds) to evaluate the performance of databases PostgreSQL 9.3 and MongoDB 3.0.3 as the persistence layer databases. The data storage architecture of the 'Consulta Opinião' project was implemented in two ways: using a single server (PostgreSQL and MongoDB) and using a replicated set of servers (MongoDB). The tests performed in local processing environment—single server—used a computer with the following specifications: Intel Core i5 2.5 GHz processor, 6 GB of RAM, SATA disk 1 TB of storage, running OS Windows 8 64-bit. The tests run in a clustered environment—replicated servers—used 3 machines with the following specifications: AMD Phenom II X2 3.2 GHz processor, 4 GB of RAM, SATA disk storage 500 GB, running Operating System Windows 7 Professional 64-bit, ethernet 10/100 Mbps.

Two types of tests, insertion and reading were performed. The insertion test is to simulate the insertion of a large number of public establishment assessments through the random data generated by the computer. The read test is to list the registered public establishments in the database, and generate an average score for each property using an aggregate query data assessment, which can be considered complex.

Figures 6 and 7 shows a comparison of the times recorded in the data entry relating to public establishment assessments. Figure 6 shows the result of the insertions of evaluations using the embedded data model; and Fig. 7 shows the result of the inserts of the evaluations using the referenced data model.

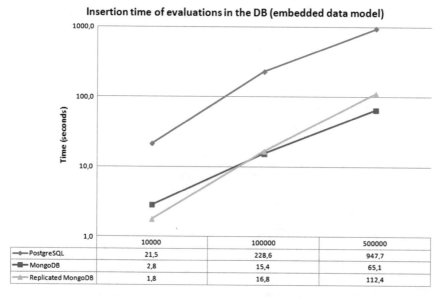

Fig. 6 Comparison insertion time of evaluations in the DB (embedded data model)

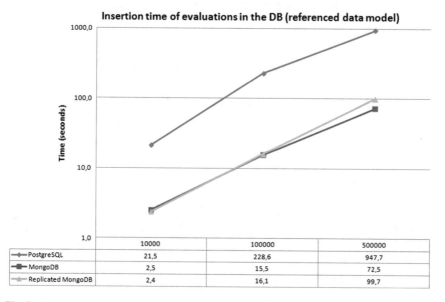

Fig. 7 Comparison insertion time of evaluations in the DB (referenced data model)

Figures 8 and 9 shows the comparison of measured times in tests of reading data stored in the PostgreSQL and MongoDB databases. Figure 8 shows the time required for reading the evaluations using the embedded data model; and Fig. 9 the time required for reading the evaluations using the referenced data model.

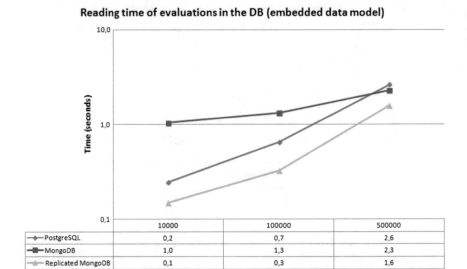

Fig. 8 Comparison reading time of evaluations in the DB (embedded data model)

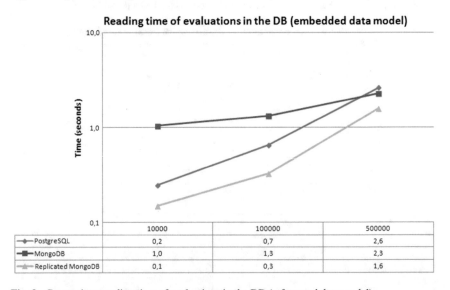

Fig. 9 Comparison reading time of evaluations in the DB (referenced data model)

Each test has its peculiarities, for example, the insertion test generates random data to make it possible to evaluate the performance of PostgreSQL and MongoDB databases in writing data. For this, there were simulations that stored, respectively, 10.000, 100.000 and 500.000 documents or public establishments assessment data. Figures 6 and 7 makes clear that MongoDB out performs the PostgreSQL in all data entry tests. By analyzing the times recorded in the data input, MongoDB achieved a performance 14 times the PostgreSQL when the number of documents passed the inserts 100.000 mark. The difference in the times recorded for data entry assessments for models embedded data (Fig. 6) and referenced (Fig. 7) it was minimum, demonstrating a similarity between the models.

On the other hand, Figs. 8 and 9 shows a better performance of the MongoDB database (replicated) on both data models used. In this case, there is also a minimum difference between the times recorded in the reading tests of models of embedded data (Fig. 8) and referenced (Fig. 9). PostgreSQL showed best performance in read operations, in relation to MongoDB with a single server, until the number of stored documents ratings exceeding approximately 400.000. From this number of stored documents, MongoDB, with single server also performs better compared with PostgreSQL. For reading tests, read operations were carried out and aggregation of data on notes generated for the evaluated properties. Reading tests were performed while the database had, respectively, 10.000, 100.000 and 500.000 registered assessments.

The results presented, MongoDB stands out mainly for its flexibility in data storage and when the amount of stored information is fairly big, for better performance in data readings compared to PostgreSQL database. The results also demonstrated that the replication of data in a cluster is a good alternative, both for insertion and for reading data.

7 Conclusions

The specification of architecture for VGIS presents diverse challenges, such as storing and manipulating large quantities of data, which, are generally structured in various formats. The NoSQL databases have flexible data schema, support various data formats and feature horizontal scalability. Therefore, NoSQL databases can be considered a good alternative for data storage VGIS.

Comparing some types of NoSQL databases it was noted that the document-based databases have greater flexibility in storage heterogeneous data, and can store files without previous knowledge of the document structure to be stored. In addition, document-based databases have good support for spatial data storage, which are a type of data to be handled by VGIS. The adoption of NoSQL MongoDB proved to be a good option because its technology supports features considered essential for VGIS as high variability of the data formats; good consistency; simplicity in the replication architecture and data fragmentation, which allows the system to adapt to work with many users and large amounts of data; in addition to support indexing and storing geographic data.

Performance tests conducted in this study, although preliminary, favored the use of NoSQL in the persistence layer of a VGIS, especially when dealing with large amounts of data.

In future works is expected to validate the storage architecture with applications that need to store files in image, audio and video format. In addition, we intend to use other NoSQL databases like CouchDB, Cassandra and Neo4j to have a basis for comparison with the results obtained with MongoDB.

References

1. Abramova, V., Bernardino, J.: Nosql databases: Mongodb vs cassandra. In: 13th International C* Conference on Computer Science and Software Engineering (C3S2E), pp. 14–22 (2013)
2. Alabri, A., Hunter, J.: Enhancing the quality and trust of citizen science data. In: IEEE Sixth International Conference on e-Science, pp. 81–88 (2010)
3. Baptista, C.S., Lima Jnior, O.F., Oliveira, M.G., Andrade, F.G., Silva, T.E., Pires, C.E.S.: Using OGC services to interoperate spatial data stored in SQL and NoSQL databases. In: XII Brazilian Symposium on Geoinformatics, pp. 61–72 (2011)
4. Brown, G., Reed, P.: Public participation IS (PPGIS) for environmental management: reflections on a decade of empirical research. URISA J. **25**(2), 5–16 (2012)
5. Câmara et al. A comparative analysis of development environments for voluntary geographical information web systems. In: XV Brazilian Symposium on Geoinformatics, pp. 130–141 (2014)
6. Camargos, B.D.C., Holanda, M., Araújo, A.: A mobile public participation geographic information system architecture for collecting opinions about public services. In: 10th Iberian Conference on Information Systems and Technologies (CISTI), pp. 1–6 (2015)
7. Castelein, W., Grus, L., Compvorts, J., Bregt, A.: A characterization of volunteered geographic information. In: 13th International Conference on Geographic Information Science (AGILE), pp. 1–10 (2010)
8. Chodorow, K.: MongoDB: The Definitive Guide. O'Reilly Media, Inc. (2013)
9. Coleman, D.J.: Volunteered geographic information in spatial data infrastructure: an early look at opportunities and constraints. In: 12th Global Spatial Data Infrastructure Association Conferences, Singapore, pp. 1–18 (2010)
10. Coleman, D.J., Georgiadou, Y., Labonte, J.: Volunteered geographic information: the nature and motivation of produsers. Int. J. Spat. Data Infrastruct. Res. **4**, 332–358 (2009)
11. Constantinides, E., Fountain, S.J.: Web 2.0: conceptual foundations and marketing issues. J. Direct, Data Digit. Mark. Pract. **9**(3), 231–244 (2008)
12. De Candia, G., et al.: Dynamo: Amazons highly available key-value store. In: 21th ACM SIGOPS Symposium on Operating Systems Principles, pp. 205–220 (2007)
13. Elmasri, R., Navathe, S.B.: Fundamentals of Database Systems. Pearson, 7 edition (2015)
14. Elwood, S.: Volunteered geographic information: future research directions motivated by critical, participatory, and feminist GIS. GeoJournal **72**(3–4), 173–183 (2008)
15. Flanagin, A.J., Metzger, M.J.: The credibility of volunteered geographic information. GeoJournal **72**(3–4), 137–148 (2008)
16. Fowler, A.: NoSQL for Dummies. Wiley, New Jersey (2015)
17. Goodchild, M.F.: Citizens as sensors: the world of volunteered geography. GeoJournal **69**(4), 211–221 (2007)
18. Goodchild, M.F.: Commentary: Whither VGI? GeoJournal **72**(3–4), 239–244 (2008)
19. Goodchild, M.F., Li, L.: Assuring the quality of volunteered geographic information. Spat. Stat. **1**, 110–120 (2012)
20. Haklay, M., Weber, P.: Openstreetmap: user-generated street maps. IEEE Pervasive Comput. **7**(4), 12–18 (2008)
21. Indrawan-Santiago, M.: Database research: are we at a crossroad?: reflection on NoSQL. In: 15th Conference Network-Based Information Systems (NBiS), pp. 45–51 (2012)

22. Jardak, C., Mahonen, P., Riihijarvi, J.: Spatial big data and wireless networks: experiences, applications, and research challenges. IEEE Netw. **28**(4), 26–31 (2014)
23. Keen, A.: The Cult of the Amateur: How Today's Internet Is Killing Our Culture. Currency (2007)
24. Kessler, F.: Volunteered geographic information: a bicycling enthusiast perspective. Cartogr. Geogr. Inf. Sci. **38**(3), 258–268 (2011)
25. Lizardo, L.E.O., Moro, M.M., Davis Júnior, C.A.: GeoNoSQL: Banco de dados geoespacial em NoSQL. Computer on the Beach, pp. 303–309 (2014)
26. Ma, Y., Haiping, W., Wang, L., Huang, B., Ranjan, R., Zomaya, A., Jie, W.: Remote sensing big data computing: challenges and opportunities. Future Gener. Comput. Syst. **51**, 47–60 (2014)
27. Maia, D.C.M., Camargos, B.D.C., Holanda, M.: Voluntary geographic information systems with document-based NoSQL databases. In: 11th Iberian Conference on Information Systems and Technologies (CISTI), vol. 1, pp. 724–729 (2016)
28. Masser, I.: Building European Spatial Data Infrastructures. Esri Press (2007)
29. McCreary, D., Kelly, A.: Making Sense of NoSQL: a Guide for Managers and the Rest of Us. Manning Publications Co., Shelter Island (2014)
30. Meek, S., Jackson, M.J., Leibovici, D.G.: A flexible framework for assessing the quality of crowdsourced data. In: AGILE International Conference on Geographic Information Science, pp. 1–7 (2014)
31. Miller, M., Medak, D., Dravzen, O.: Two-tier architecture for web mapping with NoSQL database CouchDB. In: Geoinformatics Forum, pp. 62–71 (2011)
32. Moniruzzaman, A.B.M., Hossain, S.A.: Nosql database: new era of databases for big data analytics—classification, characteristics and comparison. Int. J. Database Theory Appl. **6**(4), 1–14 (2013)
33. Nativi, S., Mazzetti, P., Santoro, M., Papeschi, F., Craglia, M., Ochiai, O.: Big data challenges in building the global earth observation system of systems. Environ. Model. Softw. **68**, 1–26 (2015)
34. Neis, P., Zielstra, D.: Recent developments and future trends in volunteered geographic information research: the case of openstreetmap. Future Internet **6**(1), 76–106 (2014)
35. Orend, K.: Analysis and classification of NoSQL databases and evaluation of their ability to replace an object-relational persistence layer. Master Thesis, Technical University of Munich, Alemanha (2010)
36. Pourabbas, E.: Geographical Information Systems: Trends and Technologies. Taylor and Francis Group, Boca Raton (2014)
37. Sadalage, P.J., Fowler, M.: NoSQL Distilled: A Brief Guide to the Emerging World of Polyglot Persistence. Addison-Wesley (2013)
38. Silberstein, A., Jianjun, C., Lomax, D., McMillan, B., et al.: Pnuts in flight: web-scale data serving at yahoo. Internet Comput. **16**(1), 13–23 (2012)
39. Sui, D.: THE 'G' IN GIS-Is neogeography hype or hope? GeoWorld **21**(3), 16 (2008)
40. Tudorica, B.G., Bucur, C.: A comparison between several NoSQL databases with comments and notes. In: 10th Roedunet International Conference (RoEduNet), pp. 1–5 (2011)
41. Vaish, G.: Getting Started with NoSQL: Your Guide to the World and Technology of NoSQL. Packt Publishing, Birmingham (2013)
42. Vera, H., Boaventura Filho, W., Holanda, M., Araújo, A.: Geographic data modeling for NoSQL document-oriented databases. GEOProcessing, pp. 63–68 (2015)
43. Vitolo, C., Elkhatib, Y., Reusser, D., Macleod, C.J.A., Buytaert, W.: Web technologies for environmental big data. Environ. Model. Softw. **63**, 185–198 (2015)
44. Zhang, X., Song, W., Liu, L.: An implementation approach to store GIS spatial data on NoSQL database. Geoinformatics, pp. 1–5 (2014)

Adding the Third Dimension to Urban Networks for Electric Mobility Simulation: An Example for the City of Porto

Diogo Santos, José Pinto, Rosaldo J.F. Rossetti and Eugénio Oliveira

Abstract Elevation data is important for precise electric vehicle simulation. However, traffic simulators are often strictly two-dimensional and do not offer the capability of modelling urban networks taking elevation into account. In particular, SUMO—Simulation of Urban Mobility, a popular microscopic traffic simulation platform, relies on urban networks previously modelled with elevation data in order to use this information during simulations. This work tackles the problem of how to add this elevation data to urban network models—in particular for the case of the Porto urban network, in Portugal. With this goal in mind, a comparison between altitude information retrieval approaches is made and a tool to annotate network models with altitude data is proposed. This paper starts by describing the methodological approach followed to develop the work, then describing and analysing its main findings, including an in-depth explanation of the proposed tool.

Keywords Intelligent vehicle software · Eco-driving and energy-efficient vehicles · Vehicle simulation systems

D. Santos (✉) · J. Pinto · R.J. Rossetti · E. Oliveira
Artificial Intelligence and Computer Science Lab, Faculty of Engineering,
Department of Informatics Engineering, University of Porto,
Rua Dr. Roberto Frias, s/n, 4200-465 Porto, Portugal
e-mail: diogo.ribeiro@fe.up.pt

J. Pinto
e-mail: carvalho.pinto@fe.up.pt

R.J. Rossetti
e-mail: rossetti@fe.up.pt

E. Oliveira
e-mail: eco@fe.up.pt

Á. Rocha and L.P. Reis (eds.), *Developments and Advances in Intelligent Systems and Applications*, Studies in Computational Intelligence 718,
DOI 10.1007/978-3-319-58965-7_14

1   Introduction

This work extends the authors' previous contribution to the *2016 11th IEEE Iberian Conference on Information Systems and Technologies* [34], by providing further context and analysis on the proposed approaches. The original contribution was motivated by a presentation with the theme "Assessing the Performance of Electric Buses: a study on the impacts of different routes" by Deborah Perrotta, based on some of the author's previous work [24, 25]. The presentation alerted to the importance of route elevation in electric vehicle simulation and the general lack of this information on many simulations, giving rise to the main issue tackled by this work: How can an elevation parameter be added to SUMO-based urban network simulations? Elevation data is important when simulating electric vehicles due to this type of vehicle's dependency on road topology, when considering its efficiency. In general, a vehicle ascending a road will put the engine under more stress than when driving in a slopeless road or in a downward direction, increasing energy consumption. In addition, electric vehicles are often equipped with energy recuperation mechanisms which can partially recharge the battery when braking or otherwise descending a road.

In particular, this problem was approached having urban networks from Porto, Portugal, as basis. Given the background of the problem, it is possible to contextualize this work in three major themes: electric vehicles, intelligent transportation systems and simulation and modelling.

The main motivation of this work is then to improve transport systems simulation (in particular in the SUMO—Simulation of Urban Mobility platform [6]) by adding elevation data to these simulations. Once completed, this project is expected to play a part in simplifying and extending the process of modelling elevation data in these simulations, simultaneously providing more accurate metrics. As a direct result, it is expected some enhancement to the precision of electric vehicle simulation in urban networks, as increased energy consumption and recuperation events related to road topology should be more precisely modelled.

In order to comply with such expectations, certain objectives were taken into consideration. The main goal considered was to find a pragmatic way of adding elevation data to strictly two-dimensional transportation systems simulations. More specifically, the work entitles some research on already existing approaches to this problem. In addition, it was set as another objective to examine the viability of generating topographic information using manually acquired GPS data. Finally, the elevation data needs to be integrated with SUMO.

The remainder of this section overviews some common issues with electric vehicles and presents to the reader a brief description of the SUMO simulator. Section 2 describes in detail the approach taken in developing this work. Section 3 details the main findings during development along with some analysis. Next, Sect. 4 makes a short description of some of the most important work found relating to this project. Lastly, Sect. 5 reviews some of the main points and contributions of this work, along with relevant future work to be developed.

1.1 Common Electric Vehicle Issues

With the use of electric vehicles, several problems arise not commonly found in their conventional counterparts.

One of the most common issues referred in the literature, especially for full electric vehicles, is the maximum range, or distance travelled with a full energy storage (be it electric batteries or hybrid counterparts) [7, 18, 38]. Authors also mention a psychological effect as consequence of the constant concern over not exceeding the maximum vehicle range, designated by "range anxiety". Alongside range issues is also the dependency on the driver profile and road characteristics such as its grade for good vehicle performance. Issues relating to road characteristics have been approached in specific by works concerning electric vehicle simulation [13].

The availability, localization and management of charging stations are also important concerns and the investigation target of many works in the literature. Yilmaz and Krein [41] mention this problem in their review, along with two other important issues, battery recharge time and the need for occasional costly battery replacements.

Yong et al. [42] mention several possible impacts of the mass adoption of electric vehicles. In particular, the need for generating more electricity may have negative impacts on power grid costs, stability and load profiles. In addition, specific environmental impacts due to the increase in power generation needs are also pointed out.

Albeit focused on the subject of *Hybrid Electric Vehicles*, Hannan et al. [12] offer further insight into challenges that pertain to the general class of electric vehicles.

As previously mentioned, the grade of the road has a critical influence on the performance of electric vehicles; in order to minimize the issues mentioned in this section, it is not advisable to disregard the road topology when planning an infrastructure that is *optimal* for electric vehicle adoption. As such, this work attempts to show the reader some options to take into account elevation data when analysing urban networks, with a particular focus on network simulations using SUMO.

1.2 SUMO—Simulation of Urban Mobility

SUMO is a traffic simulation package developed and open-sourced by the German Aerospace Center (DLR), whose development started in 2001 [2]. This tool is able to import and create urban networks for traffic simulation (vehicular and pedestrian), generate network demand models and provide routing utilities, such as shortest path calculation. SUMO is considered a microscopic traffic simulator as it is capable of detailing the behaviour of every single vehicle in the network, such as driver profiles and vehicle parameters (like mass, acceleration, etc.). The simulations are scalable, being able to represent a single traffic junction or a whole city. In addition to these features, SUMO can also be remotely configured and controlled, allowing for integrations with other tools and applications.

Using such a microscopic simulator for electric mobility simulation allows us to analyse, in detail, the performance of electric vehicles in an urban network with well defined topology, when interacting with all other kinds of vehicles, as well as pedestrians. Besides SUMO, other microscopic traffic simulation tools exist, as Passos et al. [22] show in their review, such as VISSIM [9], AIMSUN [1] and ITSUMO [36]. However, due to pre-existent work with SUMO from the authors' institution, this chapter focuses on working with that particular tool.

2   Methodological Approach

In general, the approach followed for this project started by the research and investigation on how the problem of a third dimension was solved in other work that used simulations based on SUMO. This research had the objective of analysing what kind of techniques could be used on this work. After this, it was important to select an appropriate region of Porto to use as the base model to find altitude data for. Then come phases related to retrieving the actual altitude data using any of the approaches found during the first phase, modelling that data and integrating it with the network model in SUMO. Part of the approach also involved the setup and configuration of the simulation platform, comprised of SUMO, the *High Level Architecture* and *Simulink* (these last items described in Sect. 4.1).

The following subsections describe in further detail what each of the phases of the work entailed.

2.1   Research for Possible Approaches

The main objective of this research phase was to gain further knowledge regarding simulation on the SUMO platform, how to configure and use it correctly and how the problem of three-dimensional simulation was addressed by other users. The main finding was that SUMO already supports three dimensional information to a certain degree: if this information is provided in the network file, simulations take the elevation data into consideration, including the capability of displaying in the graphical user interface information regarding inclination and height for each road segment. The main issue with support for elevation data on SUMO is that this data must be somehow provided when creating the network file, in order for the network to take that information into account. So there still remains a problem to solve: how to retrieve that information and provide it in a way that can be used by SUMO?

The following step consisted on solving this issue. Research brought to attention three possible approaches: manually capturing altitude information using a GPS device, using a combination of *OpenStreetMap* maps and data from the Shuttle Radar Topography Mission to retrieve altitude for our region of interest and, lastly, using the *Google Maps Elevation API*, by *Google*. More information about these possible approaches can be found in Sect. 2.3.

2.2 Selection of the Urban Network

Contemplating the possibility of manually capturing GPS data, it was of interest to select a region easy to visit if needed, but also with relevant elevation features to be captured, such as multiple ascents and descents of varying degrees of inclination. With this in mind, the urban network of the Porto down-town area, in particular the Aliados zone was chosen. Some previous work by Macedo et al. [13–15] contemplated the modelling of this zone into a SUMO network file ready to be loaded into the simulator, but lacking any sort of elevation data already integrated into it. As such, the present work made use of that network file.

2.3 Elevation Data Retrieval

This phase contemplated the exploration of three different approaches to retrieving elevation data. This subsection intends to define how each of these approaches can be used. In Sect. 3.1, a general comparison of pros and cons for each of these methods is given.

2.3.1 Manual GPS Information Capture

The main idea behind this approach is to use a commercial GPS device to capture GPS data, which includes altitude information. Similarly, some other attempts have tried to infer route geometry using GPS logs [10]. Since the work of this paper is related to electric vehicles—in particular, electric buses—it would be interesting to capture this GPS data along common public transportation routes in the *Aliados* network.

However, capturing a significant amount of GPS points for a good modelling of the elevation along a route can be a time-consuming task. In addition, consumer-grade GPS devices are considered to be inaccurate at determining altitude information. For both these reasons, despite being considered, this approach ended up not being used for the work, albeit being something to consider exploring in the future.

2.3.2 *OpenStreetMap* and the *Shuttle Radar Topography Mission*

The *OpenStreetMap* (OSM) project [20] is a free, editable world map being collaboratively built by volunteers, offered to the general public with an open-content license. Community efforts have made this map very accurate and detailed for many parts of the world, being a good source of information related to urban networks. As such, it is possible to import OSM map files into SUMO, in order to automatically generate accurate and detailed networks for the simulations. However, the OSM maps also lack accurate altitude information for most of the world. Nonetheless, the community has developed several tools to gather altitude information. In particular, a plugin allows linking the map to NASA's Shuttle Radar Topography Mission (SRTM) [19] data and retrieve precise altitude information for most regions of the world. The SRTM is a mission conducted since the year 2000 to obtain elevation data for most of the world with high precision and resolution and making it available to the general public.

Despite using OSM and the SRTM plugin and then directly importing the data to SUMO being simple to do and accurate regarding three-dimensional information, this approach does not allow adding the altitude value to pre-existent SUMO network files. Fully modelling an urban network to use in SUMO can be a time demanding task and disregarding old network files and starting anew in order to use an OSM file for altitude purposes may not be viable. Network files generated by OSM file conversion are often created with issues that require user maintenance to patch, such as inaccurate road representations, which translate into an added effort to use this method of network creation. As such, it is important to have an alternative to retroactively, and automatically, add the altitude information to previously existing and maintained network files.

2.3.3 Google Maps Elevation API

The Google Maps Elevation API [11] allows the retrieval of precise altitude information for any point of the world whose data *Google* possesses. The API receives (longitude, latitude) coordinates and retrieves the altitude for them. It allows querying for single points, multiple ones in bulk or along a path defined by a start and ending point.

Using this API has the advantage of directly accessing *Google's* data, which is known for its precision and availability all around the globe. However, a disadvantage of the API is related to free access limits. Free access to the API is limited to 2500 requests per day, a maximum of 512 locations per request and a total of 10 requests per second, which may make the usage of tools resorting to this API not viable for certain situations.

Nonetheless, with the objective of pre-processing and annotating SUMO network files with altitude data before running the simulations, the API was chosen to be used for the purpose of this work.

2.4 Network Altitude Model Creation and Integration of the Model with SUMO

Altitude data is combined with the original SUMO network file in order to create a new network file enhanced with three-dimensional information. These recreated files can be considered the altitude model for the urban networks, as the simulation can extract needed altitude information from it.

Since SUMO simulations rely on the network files to function, that part of the integration is set by default.

However, the developed work takes into account the possibility of, in the future, interacting with SUMO externally. As such, single point lookup capabilities were implemented and questions addressing how to store data for future queries were thought about. Conclusions regarding future work related to this issue are described in Sect. 5. Although the possibilities mentioned in the future work are not yet implemented, the development was made with future extensibility in mind. Figure 1 proposes an architecture for integrating the elevation retrieval tool, described in Sect. 3.2, using the *High Level Architecture* (HLA) approach proposed by Macedo et al. [13], described with more detail in Sect. 4.1. While a complex approach such

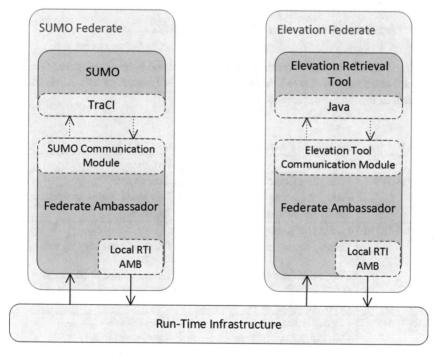

Fig. 1 Integrating the elevation retrieval tool with SUMO using HLA, based on the work of Macedo et al. [13]

as HLA may not be justified solely for integrating the tool with SUMO, when integrating multiple simulator typologies (e.g. *MATLAB Simulink* together with SUMO) that simultaneously need the tool, such an architecture may prove useful.

2.5   Urban Network Simulation

To allow for future testing and integration of the altitude data into more complex simulation systems, the simulation architecture proposed by Macedo et al. [13, 14] was installed and configured. This architecture comprises in the interconnection of the SUMO simulator with *Simulink* models, through a High Level Architecture (in the case of this work, *Pitch pRTI* [26]).

The configured system seemed to run properly after set-up. However, since the simulations did not implement usage of altitude information from SUMO yet, no testing of its performance was made.

3   Results and Analysis

3.1   General Comparison Between Methods

Section 2 mentioned that several elevation data retrieval approaches were taken into account and analysed. Table 1 summarizes the main advantages and disadvantages for each of the methods, found during research.

Table 1   General comparison of elevation retrieval approaches

	Pros	Cons
Manual GPS	+ Keep data for offline access;	- Time consuming;
	+ Adapt data to region of interest;	- May need complex manipulation;
	+ Possible path reconstruction.	- Inaccurate
OSM + SRTM	+ OSM is free, open content;	
	+ Highly detailed networks;	- Cannot integrate directly with SUMO;
	+ Can extract just a region of the world;	- Cannot be used to extend old network files
	+ Fast processing for big networks	
Google API	+ Accurate data;	- Limited free access;
	+ Near-global availability;	- Slow for real time;
	+ Ease of use.	- No offline access

Fig. 2 Summary of the tool's execution flow

3.2 The Elevation Data Retrieval Tool

The main practical output of this work was the development of a tool to automatically retrieve altitude data given coordinates in SUMO's projection space. Specifically, this tool is able to either read a previously created SUMO network file and create a new one, annotated with altitude data, or receive a single two-dimensional *(x, y)* coordinate and retrieve the corresponding elevation data.

The architecture of the tool is simple and modular. The parser/writer module parses the original SUMO network files in order to extract coordinate information and is responsible to recreate the file with altitude information. A geographic projection converter manages the conversion of coordinates in the network files to usable latitude and longitude projections. A basic *http* module queries the Google API for the needed altitude information. Lastly, our work interacts with SUMO's native *NET-CONVERT* tool to merge the altitude data to the network file, in the correct format. At the time of writing, the execution flow of the tool for network file handling, in general, is as follows, summarized in Fig. 2:

1. Parse the SUMO XML network file to extract the two-dimensional (x, y) coordinates of the relevant nodes, as well as information about the geometric projection used by SUMO (in particular, the "offsets" for the 'x' and 'y' coordinates [5]);
2. Convert each of the obtained coordinates to a latitude and longitude based projection. Currently, the tool supports converting from *Universal Transverse Mercator* (UTM) coordinates [17], as it is the projection found in the network files used for testing during development. The *Java* library *Proj4J* [21] is used to implement the geographic projection conversions;
3. Each one of the converted points is bundled into a single query to the *Google Maps Elevation API*, mentioned in Sect. 2.3.3, which retrieves the elevation data for every point. A single query is performed instead of a query per point, in order to save on the number of API calls executed for each converted file;
4. A new, temporary, XML network file is created, with the same information as the original file, but with a 'z' dimension added to the previously two-dimensional nodes, containing the elevation data;
5. Lastly, SUMO's *NETCONVERT* tool [4] is used on the aforementioned temporary network file. *NETCONVERT* uses the three-dimensionally annotated node information to generate a new network file with all the expected elevation data.

Figure 3 shows a comparison of the SUMO GUI without any kind of three-dimensional data and with provided elevation data, colouring the network edges relatively to their inclination ratio.

(a) SUMO GUI with strictly 2D informa-   (b) SUMO GUI with 3D information, dis-
tion   playing route inclination data

Fig. 3 Comparison of the SUMO GUI with and without elevation data

Table 2 Google Maps
Elevation API access times

Number of points	Round trip time (ms)
1	730
100	710
150	780
200	725
250	777
300	738

It is also possible to use the tool to retrieve data for a single coordinate, without providing any SUMO network files. In this case, the user needs only provide a UMT coordinate and the corresponding "offsets" for the 'x' and 'y' coordinates. The execution flow is similar to the aforementioned one, but in this case only the coordinate conversion step is performed before querying the API with the single point. Then, the retrieved elevation data is displayed to the user in the system console. Such an approach can be useful to extend multi-paradigm, multi-resolution, and multi-perspective simulation, integrating different models as proposed by other literature [23, 29, 30, 40].

3.3 Google Maps Elevation API access times

In order to explore a possible real-time usage of the developed tool integrated with a SUMO simulation, the average API access times, over 10 attempts, for different number of points were measured. Real-time usage could be interesting to explore in order to, instead of previously annotating the complete network file, query for the information as needed, during the simulation. The results are shown in Table 2. It is possible to observe that the access times remain approximately constant regardless of the number of points sent in the query. However, these access times are close to 0.8 s, which can be considered slow.

4   Related Work

This section summarizes two of the main works this paper is based on.

4.1   A HLA-Based Multi-resolution Approach to Simulating Electric Vehicles in Simulink and SUMO

This work defines a High Level Architecture (HLA) to interconnect different simulation systems [13]. In specific, the HLA integrates a Matlab/Simulink [39] model of an electric bus subsystem with a microscopic traffic simulation using SUMO.

Using the referred system it is possible to make a real-time analysis of electric vehicle performance under dynamic conditions. The work was tested on a scenario based in the Porto *Aliados* bus network, in Portugal. This is the same urban network used for the experimentations on the present work.

Despite the importance of elevation data for precise electric vehicle simulation, it does not take into consideration this dimension. The search for a way to correct that shortcoming motivated this work.

Besides using the same urban network for the simulations, the developments of this work were implemented with recourse to a similar system, in particular the *Simulink* + SUMO over HLA approach. This paper was found to be relevant to the present work due to the implemented HLA system. Currently, the developed tool mentioned in Sect. 3.2 does not make use of the system's functionality as it interacts with SUMO on a network file pre-processing basis. However, future work contemplates the usage of the tool as a real-time standalone altitude retrieval, to communicate both with SUMO and *Simulink* to provide altitude information when needed (and not already available on the SUMO simulation). If successful, such an improvement could be considered an important complement to the work of Macedo et al. [13], providing a more robust electric vehicle simulation platform.

4.2   Electric Vehicle Simulator for Energy Consumption Studies in Electric Mobility Systems

In this work [16], an electric vehicle model implemented directly in SUMO is suggested, as opposed to the *Simulink* model mentioned in the previously. At the time of writing of that paper, SUMO was lacking a direct way to integrate electric vehicles in its simulation capabilities. As such, the paper proposed an extension to the simulator's capabilities by adding electric vehicle models, validated with a series of tests. It is worth noting, though, that as of version 0.24, released in 2015, SUMO offers native electric vehicle simulation capabilities, along with electric charging stations data [3]. As of yet, advantages of using SUMO's native capabilities in place of the approach proposed by R. Maia et al. remain to be analysed by the authors. Despite the aforementioned electric vehicle simulation capabilities in the current SUMO versions, the paper was still found to be relevant regarding the addition of elevation data in the simulations. In particular, it proposes the integration of an independent altitude information file with the SUMO network using SUMO's *NETCONVERT* tool, which was already mentioned in Sect. 3.2. The work showed that SUMO already provided some native capabilities to integrate elevation data and motivated the usage of SUMO's NETCONVERT for the tool mentioned in Sect. 3.2. It is worth mentioning, however, that this work did not specify how that altitude information was obtained, how the file was structured nor exactly what kind of integration *NETCONVERT* did with the file. In contrast to the work developed by R. Maia et al., the current work and the developed tool intended to offer a more extensive explanation of how elevation data is obtained and integrated into SUMO, as well as provide a way to retroactively extend SUMO network files with altitude data, without any pre-prepared altitude information files.

5   Conclusions

Relating to the compared elevation data retrieval approaches, it was possible to see that all three have their advantages and disadvantages. In particular, the *Google Maps API* can be considered fast to process large bulks of data, however no clear advantage in API call time is obtained by querying for a single point. Since the average call time reaches near 0.8 s, this approach may not be suitable for real-time usage. However, the data obtained using the API is the same used in *Google Earth/Google Maps* which, in general, means good quality data. In addition, the data availability is high, with most of the world being mapped and automatic data interpolation being made in cases of unavailable points.

Regarding the *OpenStreetMap* integrated with the Shuttle Radar Topography Mission, one can consider it to be a good option to retrieve altitude data, especially because the SRTM data is very accurate. However, in order to use this approach, it is necessary to start by extracting a mapped region from *OpenStreetMap*, retrieve the altitude data using SRTM and only then generate the SUMO network file. This means that this approach cannot annotate previously existing network files. Since these files can be time-consuming to create, disregarding previous work in order to start anew using OSM + SRTM may not be viable. In addition, real time usage is not directly possible.

Lastly, when considering manual GPS altitude data capturing, one must take into account that consumer-grade GPS devices are considered to be inaccurate when acquiring altitude information. In addition, to minimize the necessity of interpolating data, which leads to even further inaccuracy, it is necessary to capture a large number of points. Since this capture needs to be done in-site, it can be a very time-consuming task. Nonetheless, keeping a local database of GPS data minimizes the need for online tools, which reduces access times and reliability issues, in turn making it a great alternative for real-time usage.

Regarding the possibility of using a local geospatial database, some considerations have been made. It has been considered integrating one of these databases within the developed tool, in order to store captured GPS coordinates. Besides the previously mentioned offline access capability, it could be possible to reconstruct travelled tracks and integrate that functionality with SUMO networks. In addition, information from other altitude data retrieval approaches could be stored as well, meaning online API accesses would only need to be made once per point and possibly pre-loaded before SUMO simulations, to allow for real-time usage.

At the moment, an initial version of the proposed tool has been implemented, capable of annotating a SUMO network file with altitude information retrieved from the *Google* Maps Elevation API. The resulting file is directly usable in any future SUMO simulations. As it stands, it is also possible to use the tool to perform altitude queries for single points (as opposed to annotating an entire network file). The tool should be open-sourced and made publicly available soon, integrated in the *MAS-Ter Lab* specification [31–33]. As main contributions from the developed work, the general comparison between altitude information retrieval techniques and the developed tool can be considered. Hopefully, these can help improve the precision of urban network simulation, in particular for electric vehicle simulation cases. However, the tool has other usages in arguably different environments, such as those using serious games and *gamification* strategies as modelling support tools [28, 37], as these can benefit from a more realistic representation of the network.

Regarding possible future work, the main priority would be implementing the local geospatial database, possibly using the *PostGIS* [27] technology. Using this database, the information retrieved from the *Google Maps Elevation API* could be stored for future offline usage and, additionally, implement some of the previously mentioned capabilities. In particular, *PostGIS* was considered for its data interpolation and track information storage capabilities "out of the box".

Since the writing of the original version of this work [34], a project from the authors' institution [35] implemented a *PostGIS*-based approach for elevation data retrieval, albeit not in direct integration with SUMO. A study on the integration of electric vehicles in a conventional bus fleet was performed in the project. To evaluate fuel and energy consumption, as well as pollutant emissions, there was a need for accurate elevation data. However, the dataset consisted on a series of two-dimensional geographic points, with no altitude information. As a solution, elevation geodata for Portugal was retrieved from the *Eurostat* database [8], converted to a *PostGIS*-compatible format and added to a *PostGIS* database. Using this geodata, a set of SQL queries was able to add elevation data to the geographic points. With such a database loaded with the adequate geodata, the data retrieval tool could use it as a source for elevation data. This would eliminate the need for an external API by integrating with an *in-house* solution.

In addition, it would be important to try and implement a *High Level Architecture Federate* for the developed tool in order to allow real-time usage in conjunction with SUMO, Further experimentation would then be needed to attest the viability of this. Lastly, some direct improvements to the developed tool's code could be made. In particular, some of the geographic projection parameters are "hard-coded" when they could be extracted from the SUMO network file.

References

1. Barceló, J.: Microscopic traffic simulation: a tool for the analysis and assessment of ITS systems. In: Highway Capacity Committee, Half Year Meeting. http://www.tss-bcn.com
2. Behrisch, M., Bieker, L., Erdmann, J., Krajzewicz, D.: SUMO—simulation of urban mobility: an overview. In: Proceedings of SIMUL 2011, The Third International Conference on Advances in System Simulation. ThinkMind (2011)
3. DLR—Institute of Transportation Systems: Models/Electric—SUMO—Simulation of Urban Mobility. http://www.sumo.dlr.de/userdoc/Models/Electric.html (2015)
4. DLR—Institute of Transportation Systems: NETCONVERT—Sumo. http://sumo.dlr.de/wiki/NETCONVERT (2015)
5. DLR—Institute of Transportation Systems: Networks/SUMO Road Networks—Sumo. http://sumo.dlr.de/wiki/Networks/SUMO_Road_Networks#Coordinates_and_alignment (2015)
6. DLR—Institute of Transportation Systems: SUMO—Simulation of Urban MObility. http://www.dlr.de/ts/en/desktopdefault.aspx/tabid-9883/16931_read-41000/ (2015)
7. Egbue, O., Long, S.: Barriers to widespread adoption of electric vehicles: an analysis of consumer attitudes and perceptions. Energy Policy 48, 717–729 (2012). http://www.sciencedirect.com/science/article/pii/S0301421512005162
8. European Commission—Eurostat: EU DEM (DD)—Eurostat. http://ec.europa.eu/eurostat/web/gisco/geodata/reference-data/elevation/eu-dem-dd (2016) (Accessed 23 June 2016)
9. Fellendorf, M.: VISSIM: a microscopic simulation tool to evaluate actuated signal control including bus priority. In: 64th Institute of Transportation Engineers Annual Meeting, pp. 1–9. Springer
10. Freitas, T.R.M., Coelho, A., Rossetti, R.J.F.: Correcting routing information through GPS data processing. In: 2010 13th International IEEE Conference on Intelligent Transportation Systems (ITSC), pp. 706–711 (Sep 2010)
11. Google: The Google Maps Elevation API | Google Maps Elevation API | Google Developers. https://developers.google.com/maps/documentation/elevation/intro
12. Hannan, M.A., Azidin, F.A., Mohamed, A.: Hybrid electric vehicles and their challenges: a review. Renew. Sustain. Energy Rev. 29, 135–150 (2014). http://www.sciencedirect.com/science/article/pii/S1364032113006370
13. Macedo, J., Kokkinogenis, Z., Soares, G., Perrotta, D., Rossetti, R.J.F.: A HLA-based multiresolution approach to simulating electric vehicles in simulink and SUMO. In: 2013 16th International IEEE Conference on Intelligent Transportation Systems—(ITSC), pp. 2367–2372
14. Macedo, J., Guilherme, S., Kokkinogenis, Z., Perrota, D., Rossetti, R.J.F.: A framework for electric bus powertrain simulation in urban mobility settings: coupling SUMO with a matlab/simulink nanoscopic model. In: 1st SUMO User Conference 2013, pp. 95 – 102. Deutsches Zentrum für Luft- und Raumfahrt e.V. Institut für Verkehrssystemtechnik Rutherfordstraße 2, 12489 Berlin-Adlershof (2013)
15. Macedo, J.L.P.: An Integrated Framework for Multi-paradigm Traffic Simulation (Apr 2013). http://repositorio-aberto.up.pt/handle/10216/72541
16. Maia, R., Silva, M., Araujo, R., Nunes, U.: Electric vehicle simulator for energy consumption studies in electric mobility systems. In: 2011 IEEE Forum on Integrated and Sustainable Transportation Systems, pp. 227–232. IEEE (Jun 2011)
17. MapTools: More details about the UTM coordinate system. https://www.maptools.com/tutorials/utm/details (2016)
18. Mehar, S., Zeadally, S., Remy, G., Senouci, S.M.: Sustainable transportation management system for a fleet of electric vehicles. IEEE Trans. Intell. Transp. Syst. 16(3), 1401–1414 (2015)
19. NASA Jet Propulsion Laboratory: Shuttle Radar Topography Mission. http://www2.jpl.nasa.gov/srtm/index.html
20. OpenStreetMap Foundation: OpenStreetMap. https://www.openstreetmap.org/#map=18/41.14987/-8.60989&layers=T
21. OSGeo Foundation: Proj4J. https://trac.osgeo.org/proj4j/ (2015)

22. Passos, L.S., Rossetti, R.J.F., Kokkinogenis, Z.: Towards the next-generation traffic simulation tools: a first appraisal. In: 2011 6th Iberian Conference on Information Systems and Technologies (CISTI), pp. 1–6
23. Pereira, J.L., Rossetti, R.J.F.: An integrated architecture for autonomous vehicles simulation. In: Proceedings of the 27th Annual ACM Symposium on Applied Computing, pp. 286–292. ACM (2012)
24. Perrotta, D., Macedo, J.L., Rossetti, R.J.F., de Sousa, J.F., Kokkinogenis, Z., Ribeiro, B., Afonso, J.L.: Route planning for electric buses: a case study in oporto. Procedia Soc. Behav. Sci. **111**, 1004–1014 (2014)
25. Perrotta, D., Ribeiro, B., Rossetti, R.J.F., Afonso, J.L.: On the potential of regenerative braking of electric buses as a function of their itinerary. Procedia Soc. Behav. Sci. **54**, 1156 – 1167 (2012). http://www.sciencedirect.com/science/article/pii/S1877042812042929
26. Pitch Technologies: Pitch pRTI Free. http://www.pitch.se/downloads/pitch-prti-free (2015)
27. PostGIS Project Steering Committee: PostGIS—Spatial and Geographic Objects for PostgreSQL. http://postgis.net/
28. Rossetti, R.J.F., Almeida, J.E., Kokkinogenis, Z., Gonçalves, J.: Playing transportation seriously: applications of serious games to artificial transportation systems. IEEE Intell. Syst. **28**(4), 107–112 (2013)
29. Rossetti, R.J.F., Bampi, S.: A software environment to integrate urban traffic simulation tasks. J. Geogr. Inf. Decis. Anal. **3**(1), 56–63 (1999)
30. Rossetti, R.J.F., Ferreira, P.A.F., Braga, R.A.M., Oliveira, E.C.: Towards an artificial traffic control system. In: 2008 11th International IEEE Conference on Intelligent Transportation Systems, pp. 14–19. IEEE (2008)
31. Rossetti, R.J.F., Liu, R.: An agent-based approach to assess drivers' interaction with pre-trip information systems. J. Intell. Transp. Syst. **9**(1), 1–10 (2005)
32. Rossetti, R.J.F., Liu, R., Cybis, H.B.B., Bampi, S.: A multi-agent demand model. In: Proceedings of the 13th Mini-Euro Conference and The 9th Meeting of the Euro Working Group Transportation, pp. 193–198 (2002)
33. Rossetti, R.J.F., Oliveira, E.C., Bazzan, A.L.C.: Towards a specification of a framework for sustainable transportation analysis. In: 13th Portuguese Conference on Artificial Intelligence, EPIA, Guimarães, Portugal, pp. 179–190. APPIA (2007)
34. Santos, D., Pinto, J., Rossetti, R.J.F., Oliveira, E.: Three dimensional modelling of porto's network for electric mobility simulation. In: 2016 11th Iberian Conference on Information Systems and Technologies (CISTI), pp. 1–6 (Jun 2016)
35. Santos, D., Kokkinogenis, Z., de Sousa, J.F., Perrotta, D., Rossetti, R.J.F.: Towards integrating electric buses in conventional bus fleets. In: 19th International IEEE Conference on Intelligent Transportation Systems (ITSC) (Nov 2016) (To be published)
36. da Silva, B.C., Bazzan, A.L.C., Andriotti, G.K., Lopes, F., de Oliveira, D.: ITSUMO: An Intelligent Transportation System for Urban Mobility, pp. 224–235. Springer (2004)
37. Silva, J.F., Almeida, J.E., Rossetti, R.J.F., Coelho, A.L.: Gamifying evacuation drills. In: 2013 8th Iberian Conference on Information Systems and Technologies (CISTI), pp. 1–6. IEEE (2013)
38. Steinhilber, S., Wells, P., Thankappan, S.: Socio-technical inertia: understanding the barriers to electric vehicles. Energy Policy **60**, 531–539 (2013). http://www.sciencedirect.com/science/article/pii/S0301421513003303
39. The MathWorks Inc.: Simulink—Simulation and Model-Based Design. https://www.mathworks.com/products/simulink/ (2015)
40. Timoteo, I.J., Araujo, M.R., Rossetti, R.J.F., Oliveira, E.C.: TraSMAPI: an API oriented towards multi-agent systems real-time interaction with multiple traffic simulators. In: 13th International IEEE Conference on Intelligent Transportation Systems, pp. 1183–1188. IEEE (Sep 2010)
41. Yilmaz, M., Krein, P.T.: Review of battery charger topologies, charging power levels, and infrastructure for plug-in electric and hybrid vehicles. IEEE Trans. Power Electron. **28**(5), 2151–2169 (2013)

42. Yong, J.Y., Ramachandaramurthy, V.K., Tan, K.M., Mithulananthan, N.: A review on the state-of-the-art technologies of electric vehicle, its impacts and prospects. Renew. Sustain. Energy Rev. **49**, 365–385 (2015). http://www.sciencedirect.com/science/article/pii/ S1364032115004001

Parallel Remote Servers for Scientific Computing Over the Web: Random Polygons Inscribed in a Circle

Marco Cunha and Miguel Casquilho

Abstract Computing over the Web can be applied to solve problems of many types of users, namely companies, which can thus benefit from the service of remote collaborators, e.g., from academia, without the user's awareness, namely if the user is the customer. An illustration of scientific computing over the Web is described, based on remote servers in parallel, as fits separable time consuming tasks. The illustrative problem chosen was the probabilistic behaviour of the perimeter of random polygons inscribed in a circle, because: it is simple to state; its Monte Carlo resolution suits the type of computing considered; and each polygon to be simulated (with 3, 4, etc., sides) can be assigned to multiple servers, only two servers for demonstration purposes. Our objectives are: to emphasize the pertinence of the Web for scientific or other computing; and to show the use of parallel, possibly heterogeneous remote servers. The problem shown can be freely solved on the authors' webpage. We think the benefits of scientific computing over the Web, despite the ubiquity of this environment, have lagged behind most of the other uses of the Internet.

Keywords Scientific computing · Web · Parallel remote servers · Monte Carlo

1 Introduction

In our meaning of "scientific computing over the Web", a program is available on the Web for a user to run it in order to solve a problem just using a Web browser. No software installation is supposed on the user's terminal, and the user can be anonymous or identified, through authentication, the latter case being common in

M. Cunha · M. Casquilho (✉)
IST, Universidade de Lisboa (University of Lisbon), Ave. Rovisco Pais, IST,
1049-001 Lisboa, Portugal
e-mail: mcasquilho@tecnico.ulisboa.pt

M. Cunha
e-mail: marco.cunha@tecnico.ulisboa.pt

© Springer International Publishing AG 2018
Á. Rocha and L.P. Reis (eds.), *Developments and Advances in Intelligent Systems and Applications*, Studies in Computational Intelligence 718,
DOI 10.1007/978-3-319-58965-7_15

the context of a company. Also, in research activities, computing over the Web can make sharing assignments easier among geographically distributed members. We address the case where several, possibly heterogeneous, remote servers are used in parallel, and this has the advantage of being transparent, i.e., without the user's unnecessary awareness, to solve a problem. For instance, a company might have purchased a Windows software for a certain task and a Linux one for another task, and so on, each in different subsidiaries, and now it would be necessary to combine it all for a new purpose to meet the customer's needs.

In the academic practice of the second author, many programs were, since more than two decades, written and made available on our webpages to students of Operational Research, Quality Control, Engineering, and Computing. The structure of the computing system can be kept visible to the students for obvious pedagogical reasons; notwithstanding, namely, in the context of companies, it is not pertinent or necessary to reveal the technical implementation details to the customers, all the more so in the many cases when the computing is outsourced to real time remote systems. Indeed, the only preoccupation of the customers is the correctness of the final results.

While the work on one's own computer (not over the Web) depends on installing some sort of software, the "Internet era" allows different approaches. Installing software implies: adequacy of the software to the user's platform (operating system) and its power and capacity (hardware, operating system, speed, memory); maintenance (security, updates, upgrades); the necessary permissions to the software for each user; and eventual uninstallation if the licence expires or the software is not of interest, with its issues, such as the frequent registration debris in the Windows systems and the remaining temporary files.

We have addressed scientific computing over the Web in several previous works (e.g., [1, 2]), besides our daily use in teaching. We talk about scientific computing, but of course any type of computing (e.g., [3]) over (or on) the Web may be envisaged.

The question of scientific computing itself is not its relative absence on the Web. Indeed, innumerable websites offer (frequently free) software for computing, scientific or not, but relatively few websites provide scientific computing over the Web, therefore, specific references to this type of computing are not easy to find (or so they were not for us). An outstanding exception (incidentally, in the same tone as ours) is Ponce's [4] extensive library in Hydraulics and Hydrology, written in Fortran, although lacking systematic default user data, perhaps exactly for pedagogical reasons. His website, however, to our knowledge, uses no parallelism to perform the calculations, which increases the execution time.

Many websites provide scientific computing, but with one or more aggravations, among others: Java required (with its stringent security rules since version 7) and its incompatibility with several popular browsers, such as Google Chrome or Safari; lack of user data customization (namely, numerical data); and browser plugins needed such as Adobe Flash. A number of websites provide vast information on Mathematics and Computing, such as Wolfram MathWorld [5] and its WolframAlpha [6] using their own *Mathematica* language. Other websites are based on (invisible) advanced computing, such as the Google search engine, but without computation as a direct objective. Several other entities (e.g., [7, 8]) provide useful resources, although targeting various purposes that are different from ours, such as cloud storage and "big data" computing.

The scientific problem that underlies the present study, the probabilistic behaviour of the perimeter of random polygons inscribed in a circle, which is just a theoretical exercise, was devised in order to feature: distinct (and complementary), parallelizable computing tasks; an abstract application without any association to a specific situation, thus not obscuring the computing aspects central to this text; and a Monte Carlo resolution technique, suitably general and well-known in the scientific and technical community, allowing (almost) for arbitrarily large time consuming and independent runs, a characteristic that is suitable to the adoption of parallel computing.

The problem in this study differs considerably from a previous one [9] presented by the same authors. Only the general technique of Monte Carlo simulation is common to both cases. The previous work used remote *serial* computing, even though in remote machines, for statistical sampling, whereas the present work is about a completely different mathematical problem (random polygons), run in *parallel*, able to run remotely in possibly different environments (operating systems, geographical sites, programs) and without any third party applications.

The problem addressed lends itself to distribution to various servers, ideally in parallel. Whether the servers are remote or not might be irrelevant, but remoteness is in the objective of this study. In the current situation, the fault tolerance issue is not explored because the computation strictly depends on each task performed by each remote server.

Monte Carlo simulations are "brute force", extensive calculations (the more, the better) that give a response to a problem for which there is no analytical solution or it is not known or practical, as (to our knowledge) happens in our illustrative case. The problem, as we state it —run on possibly totally different computers—, should be solved neither in purely OpenMP [10] nor in MPI [11]. The distribution to two (or more) remote servers presupposes that each of the non-interchangeable servers may have to provide distinct, possibly proprietary software or other resources that may be specific to each server.

The computing model is based on a client-server Web system paradigm, requiring relatively few, negligible resources from the client, and assigning the (arbitrarily large) computing effort to the servers. Indeed, the only resources needed in the client computer are a Web browser and a connection to the Internet, since all the sizable computational tasks are performed at the remote servers.

The computing: is launched from a public (or, optionally, password protected) webpage, the *Expose* at system *Local*; is done at several (two, in our study) remote servers, the *Assistant*'s at system(s) *Remote*; and is finally returned to the *Conclude* at system *Local* to show a webpage with the results. The system *Local* is, in this study, a virtual machine that was granted us by CIIST (our School's "Informatics Centre"), a machine that is independent from the main School system (the engineering school of the University of Lisbon), i.e.: *Local* is at *sasws.tecnico.ulisboa*.

Table 1 Programs, their system identification, and address

Program	System	Address
Expose	Local	sasws.tecnico.ulisboa.pt
Assistant 1	Remote 1	web.tecnico.ulisboa.pt
Assistant 2	Remote 2	web.tecnico.ulisboa.pt
Conclude	Local	sasws.tecnico.ulisboa.pt

Table 2 Programs, their system identification, and (as examples) their address and operating systems (OS)

Program	System	Address	OS
Expose	Local	our.company.com	Linux
Assistant 1	Remote 1	my_desktop_pc	Windows
Assistant 2	Remote 2	some.school.system	Linux
...	...	...	...
Assistant n	Remote n	our_affiliate	Mac
Conclude	Local	sasws.tecnico.ulisboa.pt	Linux

pt; and the *Remote* s are at our School web system, at *web.tecnico.ulisboa.pt*. (In our previous related study [9], a geographically fully separate source account was hosted at www.gsd.inesc-id.pt, a Linux system. A secure shell connection, however, could not be timely updated so as to be used in this succeeding study). The systems and programs are shown in a clearer way in Table 1.

Incidentally and without loss of generality, in the present study, *Remote 1* and *Remote 2* are at the same address, as shown in Table 1. Let it be noted that this address corresponds to a cluster of 5 servers, thus each job is automatically assigned to one of these servers, according to a typical load balance algorithm, making it probable that the tasks be really executed by physically autonomous machines.

The structure in Table 1 is a particular case of the general structure shown in Table 2. In this latter table, the major feature is that the servers are heterogeneous, running different operating systems (OS).

While in Table 2 the heterogeneous systems are clearly stated, our illustration in this study deals with "pseudo-heterogeneous" servers. As a matter of fact, we did not have available such a set of different computers, so we did a similar scheme with various Linux accounts. Obviously, the same structure might have been used with other hardware and operating systems, as exemplified in the already mentioned Table 2.

The programming language chosen for the mathematical computation was Fortran 90, adequate to "scientific" computing. Another language, such as C, might be chosen (or combinations of these or more), as our goal is to implement communication between a webpage and two (or other number of) remote servers. No scientific argument, otherwise, reproves other languages, such as Matlab or Mathematica, for the current purpose, but their licencing does not facilitate their use on the Internet. An alternative for the *Assistant*'s doing these computations might be to use grid computing, as suggested in [7].

In sum, in this study on scientific computing over the Web, we intend to: stress the use of the Web for this purpose; exemplify this type of computing with a time consuming task, as is typical of the Monte Carlo applications; use parallel remote servers, possibly heterogeneous; and once more, taking advantage of the Web, remind a way to ease links, such as the one between the university and industry.

In the following sections: the illustrative mathematical problem and its method of resolution are described; the computing strategy and results are shown and discussed; and some conclusions are presented. The problem can be freely run in the authors' website [12] constructed for this study.

2 Perimeter of Random Polygons

Obtaining the perimeter of random polygons inscribed in a circle will mean here to find its empirical probability density function, as follows. This kind of problem is a typical application of the Monte Carlo method. Suppose that a polygon, such as a triangle or a quadrilateral or a pentagon, etc., with $n = 3, 4, 5$, etc., respectively, has its n vertices randomly located on a circle, defining its n sides. We assume, for simplicity, that: the polygons (obviously from $n = 4$ onward) are not starred; and, without loss of generality, the circle has unit radius, $R = 1$. (For other radii, the perimeter is of course proportional.) In Fig. 1 are shown two triangles, a larger and a smaller one, this latter to remind that the perimeter can be near zero or even zero (degenerate triangle), if the three vertices happen to coincide. In Fig. 2 are shown two quadrilaterals, one being "legitimate" and another "illegitimate", because it is starred, a case not addressed in this study. Of course, the latter case (starred polygon) can be "corrected", provided that, as was done, the angles of the four vertices will be sorted (ascendingly or descendingly). To this end, a simple bubble

Fig. 1 Triangles inscribed in a circle

Fig. 2 A non-starred
quadrilateral and a starred
(illegitimate) quadrilateral
inscribed in a circle

sort algorithm was used, no more efficient algorithm being (in our opinion) necessary because at most only nine vertices (enneagon) are simulated.

The perimeter of a random polygon with n vertices in a circle of radius R is
limited above by the perimeter of the corresponding regular polygon [13], as given
in Eq. (1).

$$P = 2nR \sin\left(\frac{\pi}{n}\right) \tag{1}$$

In order to plot the probability density function, pdf, and the cumulative distribution function, cdf, to accommodate the cases for any n, the worst case (leading
to the highest value of P is (e.g., [13]), then making $R = 1$ m (or other coherent unit
of length),

$$P_{\max} = \lim_{n \to \infty} 2nR \sin\left(\frac{\pi}{n}\right) = \lim_{n \to \infty} 2n\frac{\pi}{n} = \\ = 2\pi \tag{2}$$

So, the values of P (always in the same length units) will not surpass 2π,
i.e., ~6.28, which will be used as the maximum abscissa for the plots of the
simulated 'pdf' (and 'cdf'). In the simulations done, it was assumed that n is in the
interval

$$3 \leq n \leq 9 \tag{3}$$

Indeed, the results will show two apparently distinct types of behaviour, for
$3 \leq n \leq 4$, and for $5 \leq n$, respectively: a surprising pattern for random triangles
and quadrilaterals; and, for pentagons ($n = 5$) and above, another type of graph,
showing no interesting modification of shape as n increases. This is why, to save
space, no graphs are included for the higher values of n.

For the simulation of each polygon of n vertices, the procedure was the following (with the notation $j .. k$ meaning all the integers from j to k, both extremes included):

(a) Simulate n random uniform (0–1) values, r_i, $i = 1 .. n$.
(b) Calculate the vertices' angles, θ_i, $\theta_i = 2\pi r_i$.
(c) Sort θ_i to avoid star polygons (see comment).
(d) Calculate the coordinates of each vertex, $x_i = \cos \theta_i$ and $y_i = \sin \theta_i$.
(e) Calculate the sides (distances), s_i, from each point to the subsequent one (points i to $i + 1$ for $i = 1 .. n{-}1$, and n to 1).
(f) Make a histogram of the values of the sides.

This procedure can be observed in the authors' website [14] for any particular polygon, obeying the condition in Eq. (3). The user's data include the number of Monte Carlo trials, N, the random number generator seed (iff 0, "non-repeatable" series of numbers), and the number of classes (points) for the resulting histogram. So, the problem can be solved on a single system, for a mere comparison as available on the webpage mentioned, running a Fortran 90 program, where default values are proposed.

Monte Carlo runs typically consume considerable (sometimes arbitrarily long) time, although they begin, for meaningful results with triangles, at about 1 s. So, the distribution to several servers in parallel, e.g., one for each n side polygon, appears to be the right choice, keeping in mind the supplementary fact that they are allowed to be heterogeneous and remote.

3   Execution of the Simulation

Simulation runs for triangles, quadrilaterals, pentagons and hexagons, i.e., $n = 3, 4,$ 5, and 6, are shown in Figs. 3, 4, 5 and 6, respectively, up to $n = 9$ being able to be run in the webpage. The figures have the same scales, making comparison easy. The value of $N = 20$ million (20e+6) polygons was found to be "sufficiently" large and not too lengthy, i.e., the results are stable and such simulations can be done from the webpage, not exceeding the maximum time permitted by the system's administration, about 30 s. Much lengthier runs were done from the command line, where there is no such restraint, but no further improvement in the results was observed.

The curves are just histograms built from the experimental frequencies in 200 classes. In each graph, the curve with a peak, related to the left-hand y-axis, is the 'pdf', and the rising curve, related to the right-hand y-axis, is the 'cdf', computed by numerical "integration" (summation) of the former. (The 'cdf' curve obviously goes from 0 to 1, as it is the accumulated probability.) The surprising behaviour deserves analysis in a context of Statistics, but that is out of the scope of this study on computing. Indeed, the perimeters of all the polygons always show a peak, which is sharp for triangles and quadrilaterals, whereas it looks smooth for polygons with

Fig. 3 Perimeter of random
triangles (1 s)

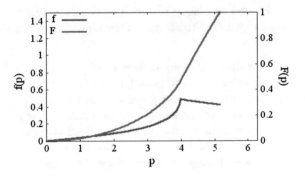

Fig. 4 Perimeter of random
quadrilaterals (2 s)

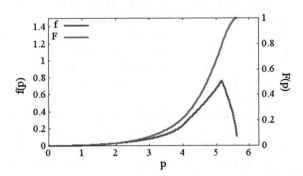

Fig. 5 Perimeter of random
pentagons (4 s)

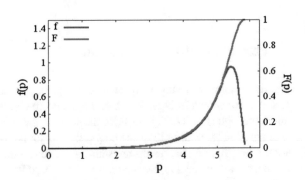

Fig. 6 Perimeter of random
hexagons (7 s)

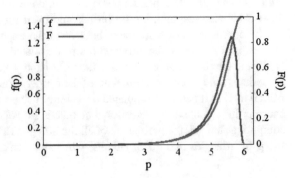

Table 3 Average, standard
deviation, and peak (mode)
coordinates from the
simulation runs

Polygon	Average, std. dev.	Mode x and y
Triangle	3.82, 1.07	4.05, 0.49
Quadrilateral	4.54, 0.86	5.22, 0.76
Pentagon	4.99, 0.68	5.51, 0.95
Hexagon	5.29, 0.55	5.69, 1.26

more sides. The computing times (always for the same N above) are indicated in the figures' captions. As a final synthesis of the results for each perimeter of the polygons tested, Table 3 shows its average and standard deviation, with the coordinates x and y of the peak (the mode).

In order to prepare much lengthier experiments for the parallel runs, command line executions were performed, avoiding the system-wide Web limit of about 30 s. The results suggest $N = 500$ million (500e6) yielding a run time of about 3 min for hexagons, the most time consuming of the few cases examined. These results are given in this section because they are predominantly statistic, whereas the aspects related to computing (information technology itself) will be discussed in the next section.

The illustration of scientific computing used may originate extensive computations. Due to its structure of independent problems, it is suitable for parallel computing that can be distributed to various (possibly different) servers, as will be shown for two remote servers.

4 Results and Discussion

The application being supposed to be used on the Web, there is no advantage that the computing should be visible to the user, even in the case where the user may be its sole "customer", as in the case of a company. The only information accessible to the user is his input (data) to the application and the output (results), both being HTML files built in the scripting language PHP. This Web-friendly language was chosen due to its intrinsic ease of integration with the Web browsers, and the availability of *out-of-the-box* libraries for our requirements: integration with system calls and SSH (secure shell server) protocol, and efficient performance.

The architecture has the following components: (i) a Web-based front-end, to offer a user interface so that the user may insert his problem input; and (ii) a computing back-end, constituted by two (or more) servers, to execute the respective tasks, one independent task per remote server. The language selected (Fortran 90) was chosen because it is: generally adequate and efficient for the numerical problems in industry; free from intricate licensing conditions, considering its target "public" use on the Web; and able to readily produce a standalone executable.

The front-end Web interface is accessed by the user, where the job's data are accepted. The front-end schedules the job and forwards the data to the back-end computing servers. The back-end, after finishing the various jobs, replies to the front-end with the gathered output, from the various (two) remote servers. In the examples given, the output is a set of numerical results used to produce a graphic, also shown. In the cases of "classical" languages (such as Fortran and C), which have no graphical capabilities, the graphic is made via a graphic tool, which, in our case, was the common free *gnuplot*. The results, i.e., the numerical output and the graphic, are formatted as HTML (via PHP) for the Web and finally presented in a way comprehensible by the user.

The architecture consists of the following. The user addresses the problem webpage and introduces his data (input) in the fields of an HTML *form* inserted in a PHP page. Clicking the 'Execute' button sends the data by the POST method to another PHP file, managing the distribution to the remote servers, by SSH connections. This PHP has as objective to establish the connection between the various programs and distribute the tasks in a parallel mode, and collect the final results, including the graphic.

The final page, visible by the user, contains the HTML *src img* tag with the reference to the graphic file (a 'png' file), which was just generated and automatically sent to a temporary directory accessible to the Web. In order to permit simultaneous (independent) executions by several users, the graphic file is created using unique names generated by a function of the wall clock time through PHP *microtime*.

The files needed are the following, with the indication of which physical system, local or remote, each is at:

(g) **remote_parallel.php** at *Local* (see Table 1)—webpage visible by the user, where: the user accesses the particular problem with a browser and the procedure begins; and the user inserts his input data, which are handled by the subsequent file.

(h) **dispatcher.php** at *Local*—file for handling the problem data received from 'remote_parallel.php' for execution by the specific executable programs, all remote, available and adequate for the particular problem. This file controls the flow, invoking: serially, the first program (*Expose*) to display the problem data; then, in parallel, the various remote programs (in our case, the two *Assistant*'s); and finally a last program (*Conclude*), synthesizing the results from all the previous computations, and producing an illustrative graphic. This file, 'dispatcher.php', uses all the output already produced, and generates dynamically an output PHP webpage with the results available to the user.

(i) **Expose.exe** at *Local*, executable program—local standalone executable, written (in this example) in Fortran, compiled and linked, which just prints (echoes) the input data of the particular problem, for confirmation of the data used in all the subsequent computations.

(j) **Assistant1.exe** and **Assistant2.exe** at *Remote1* and *Remote2*, respectively, executable programs—remote standalone executables, written (in this example) also in Fortran, compiled and linked, solving the (two) middle, independent parts of the particular problem, simultaneously (but not necessarily with the same duration). These programs return each a public URL, to be used by the subsequent program to collect the output data.

(k) **Conclude.exe** at *Local*, executable program—local standalone executable, written (in this example) also in Fortran, compiled and linked, which receives the URLs of the results from the *Assistant*'s and the URL with the settings for the graphic (to be used in the HTML tag '<img src = …>'), and produces the graphic, returning its URL.

(l) **Complementary files** at *Local*—**sshConnection.php, utils.php** (explained below), PHP environment files, cascading style sheet, and the images that characterize the website.

The file 'sshConnection.php' contains the methods necessary to establish the secure connections with the executables, and implements robust criteria to validate the connections and handle possible issues. The file 'utils.php' contains several functions: formatting the output data from the programs; creating the local file with the settings for 'gnuplot'; parsing the output from each executable to extract a URL, to both inform the last executable ('Conclude.exe') about the coordinates for the graphic, and supply the location for the graphic. (The command line program 'gnuplot', as previously mentioned, is a well-known free program used to generate two- or three-dimensional plots.)

The fact that the problem is solved "over the Web" adds no further difficulty in relation to the basic resolution of any scientific problem. Scientific computing frequently leads to batch type execution, without interactivity, so the present paradigm is adequate. The use of the Web adds the recognizable advantage of the access to many remote services that may be geographically separate.

The architecture adopted is shown in Fig. 7, with the nomenclature already introduced, the numbers (in the vertical axis) meaning: 0, the *Local* system, where the public webpage is located, i.e., where the user inserts the problem data via the program *Expose* and will receive the results in a webpage, via the program *Conclude*; 1, 2, etc., the *Remote* systems, where the various components of the computation are done.

The fact that the remote part of the computing flow is performed in parallel reduces the time effort to complete the process. In the demonstrative context of this study: there are no fault tolerance mechanisms against failure of any of the servers; and no input data validation is performed. Our results show, in accordance with

Fig. 7 Flow of computing, with numbers meaning: 0, *Local* system (with programs *Expose* and *Conclude*); 1, 2, …, *Remote* systems in parallel (with programs *Assistant*)

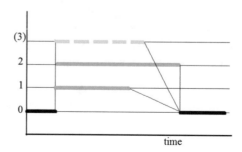

Amdahl's law [15], that, in the case of two identical runs, the total time spent was not reduced to half. By "identical runs", we mean the same polygons, e.g., triangles, and the same number of simulated perimeters. Indeed, there is a fixed workload, related to the task management carried out by the operating system, and by the managing procedure ('dispatcher.php'), as well as the communication overheads.

The problem addressed is, naturally, available for execution in the mentioned authors' website [12]. In the context of this study, the system, as a whole, is simple to create and maintain, and proved, on the usability side, to be quite practical and friendly.

5 Conclusions

The Web can provide access to scientific (or other) computing in a way that has been relatively little explored, namely in the context of companies and the much praised collaboration with academia. It makes available computing systems of any power, regardless of the capability, type or operating system of the user's terminal, just through a Web browser, without any particular software installation.

The illustration for this study is the Monte Carlo simulation of polygons (with 3 to 9 vertices), inscribed in a circle, in order to get their perimeter and assess its probabilistic behaviour. This problem clearly suggests the distribution by several servers (here, two) in parallel, one for each polygon case, and possibly different servers from each other. The Monte Carlo simulation is a powerful technique that can be arbitrarily lengthy. Although the computation was done on similar remote systems (anyway, two completely independent executions on the University computing system), the use of a system assigning tasks to remote servers is an example of a distribution that can be mandatory when the servers perform heterogeneously, with different functions, even on different platforms. The gist of this work was to show how this can be done without the user's perception, such as is adequate in a company towards its workers and in its relation with its customers.

The problem addressed permitted to show the suitability of the Web as an environment for scientific (or other) computing with the ability to use various remote homogeneous or heterogeneous servers to accomplish a given goal. The mathematical method used was a simple Monte Carlo simulation, without further statistical exploration, which would obscure the computational character of the study.

Computing over the Web uses the same executables as in the command line. During two decades of academic practice, one of the authors has verified that Web computing is adequate, whether for academia or industry purposes or for the connection—ever in need of improvement—between any of these types of entities.

Acknowledgements The study was done at CERENA (Centre for Natural Resources and the Environment), funded by FCT, *Fundação para a Ciência e Tecnologia* (Portuguese "Foundation for Science and Technology"), Project UID/ECI/04028/2013, and Department of Chemical Engineering, IST (*Instituto Superior Técnico*), *Universidade de Lisboa* (University of Lisbon),

Lisbon, Portugal. For the study, the authors thank: (M. Casquilho) the Department of Chemical Engineering, and (M. Cunha) the Department of Information Technology and Computers, both of IST; CIIST (the Computing Centre of IST); and Professor João Coelho Garcia (Department of Information Technology and Computers, IST, and Inesc-ID, Institute for Systems Engineering and Computers, Research and Development, Lisbon). The referees' comments contributed to the clarity of this report.

References

1. Ferreira, M., Casquilho, M.: Scientific computing over the Internet; an example in Geometry. In: WorldCIST'13, World Conference on Information Systems and Technologies, Olhão (Faro, Portugal) (2013)
2. Casquilho, M., Cunha, M.: Scientific computing over the Web in various programming languages—solving problems in Fortran, C, and Octave. In: CISTI'2014, 9ª Conferencia Ibérica de Sistemas y Tecnologías de Información (9th Iberian Conference on Information Systems and Technologies), Barcelona (Spain) (2014)
3. Sati, M., Vikash, V., Bijalwan, V., Kumari, P., Raj, M., Balodhi, M., Gaurila, P., Semwal, V. B.: A fault-tolerant mobile computing model based on scalable replica. Int. J. Artif. Intell. Interact. Multimed. 2(6), 58–68 (2014)
4. Ponce, V.M.: San Diego State University. http://ponce.sdsu.edu/online_calc.php (2016). Accessed 01 Feb 2016
5. Wolfram MathWorld. http://mathworld.wolfram.com/ (2016). Accessed 01 Feb 2016
6. WolframAlpha. http://www.wolframalpha.com/input/?i=3%2B9 (2016). Accessed 01 Feb 2016
7. EGI: European Grid Infrastructure. http://www.egi.eu/
8. SCI-BUS: SCIentific gateway Based User Support. https://www.sci-bus.eu/
9. Cunha, M., Casquilho, M.: Applied scientific computing over the Web with remote servers: sampling with and without replacement. In: CISTI'2015, 10ª Conf. Ibérica de Sistemas e Tecnologias de Informação (10th Iberian Conf. on Information Systems and Technologies), Águeda, Aveiro (Portugal) (2015)
10. OpenMP: The OpenMP API specification for parallel programming. http://openmp.org/ (2015). Accessed 01 Feb 2015, complete specification of July 2013
11. Argonne National Laboratory: The Message Passing Interface (MPI) standard. http://www. mcs.anl.gov/research/projects/mpi/ (2016). Accessed 01 Feb 2016
12. Casquilho, M.: Random polygons: remote parallel servers. http://sasws.ist.utl.pt/MCasquilho/ CISTI_2016/remote_parallel.php
13. eFunda.com: Regular polygon inscribed in a circle. http://www.efunda.com/math/areas/ PolygonInscribedGen.cfm (2016). Accessed 01 Feb 2016
14. Casquilho, M.: Perimeter of inscribed polygons, at Técnico (IST). http://web.tecnico.ulisboa. pt/~mcasquilho/compute/qc/Fx-polygonPeriForParal.php
15. Amdahl, G.M.: Validity of the single processor approach to achieving large-scale computing capabilities. In: AFIPS Conference Proceedings, vol. 30, pp. 483–485 (1967)

webQDA 2.0 Versus webQDA 3.0: A Comparative Study About Usability of Qualitative Data Analysis Software

António Pedro Costa, Francislê Neri de Souza, António Moreira and Dayse Neri de Souza

Abstract Communication and Information Technology (ICT) applied to qualitative research has proved to be an asset for quality, improvement and innovation of this research area. Having emerged in 2010 webQDA is an online software that has supported the data analysis of several areas of research such as education, health, sociology, psychology, social sciences and others fields that need qualitative analysis. This study presents an assessment of the usability of the qualitative data analysis software webQDA® (version 2.0) in comparison with its new version webQDA 3.0. To assess its usability, the System Usability Scale (SUS) was used. The results indicate that there is still room and potential for improvement of webQDA 3.0, as expected, once it is still being improved. It is encouraging to know that more and more researchers find it not only useful but easy to use, even if demanding (which is positive from our point of view), in the way it "forces" researchers to adopt a more inquisitive approach to the data they deal with in their qualitative studies. Suggestions for future studies and improvement of the software are put forward.

Keywords Usability · Qualitative computing · Qualitative research · Qualitative data analysis · Research through design

1 Introduction

The evaluation of usability is much discussed, especially when we approach the graphic interfaces software. The study of usability is critical because certain software applications are "one-click" away from not being used at all, nor used in an appropriate manner.

A.P. Costa (✉) · F.N. de Souza · A. Moreira · D.N. de Souza
CIDTFF - Research Centre "Didactics and Technology in Education of Trainers",
DEP/UA – Education and Psychology Department, University of Aveiro,
Aveiro, Portugal
e-mail: apcosta@ua.pt

© Springer International Publishing AG 2018
Á. Rocha and L.P. Reis (eds.), *Developments and Advances in Intelligent Systems and Applications*, Studies in Computational Intelligence 718,
DOI 10.1007/978-3-319-58965-7_16

Usability comes as the most "rational" side of a product, allowing users to reach specific objectives in an efficient and satisfactory manner. Moreover, User Experience is largely provided by the feedback on the usability of a system, reflecting the more "emotional" side of the use of a product. The experiment is related to the preferences, perceptions, emotions, beliefs, physical and psychological reactions of the user during the use of a product [1]. Thus, one comprehends "the pleasure or satisfaction" that many interfaces offer users, as evidence of efficiency in the integration of the concepts of Usability and User Experience in the development of HCI solutions [2]. Despite these two concepts being linked, in this article we will focus our attention on the dimension of Usability.

When dealing with authoring tools, in which you have to apply your knowledge to produce something, it becomes even more sensitive to gauge the usability of the software. Being webQDA® (Web Qualitative Data Analysis: http://www.webqda. net) an authoring tool for qualitative data analysis and a new version of it being developed (whose release took place April 2016) [3], it is of extreme relevance to keep this dimension in mind.

webQDA is a qualitative data analysis software that is meant to provide a collaborative, distributed environment (http://www.webqda.net) for qualitative researchers. Although there are some software packages that deal with non-numeric and unstructured data (text, image, video, etc.) in qualitative analysis, webQDA is a software directed to researchers in academic and business contexts who need to analyze qualitative data, individually or collaboratively, synchronously or asynchronously. webQDA follows the structural and theoretical design of other programs, with the difference that it provides online collaborative work in real time and a complementary service to support research [4].

In this article, the main objective is to answer the following question: How to assess usability and functionality of the qualitative analysis software webQDA?

After these initial considerations, it is important to understand the content of the following sections of this paper. Thus, Sect. 2 will present concepts associated with Qualitative Data Analysis Software. Section 3 will discuss Design through Research, while Sect. 4 will assess webQDA from the point of view of Usability, a methodological aspect where we present the methods and techniques for assessing the usability of software and the results of this study. And finally, we conclude with the study's findings.

2 Qualitative Data Analysis Software (QDAS)

The use of software for scientific research is currently very common. The spread of computational tools can be perceived through the popularization of software for quantitative and qualitative research. Nevertheless, nonspecific and mainly quantitative tools, like Word®, SPSS®, Excel®, etc., are the ones with major incidence or dissemination. This is also a reflection of a large number of books that can be found on quantitative research. However, the integration of specific software for qualitative research is a relatively marginal and more recent phenomenon.

In the context of postgraduate educational research in Brazil, some authors [5] studied the use of computational resources in research. They concluded that among those who reported using software (59.9%), there is a higher frequency of the use of quantitative analysis software (41.1%), followed by qualitative analysis software (39.4%), and finally the use of bibliographic reference software (15.5%).

The so-called Computer-Assisted Qualitative Data Analysis Software or Computer Assisted Qualitative Data Analysis (CAQDAS) are systems that go back more than three decades [6]. Today we can simply call them Qualitative Data Analysis Software (QDAS). However, even today many researchers are unaware of these specific and useful tools. Puebla [7] specifies at least three types of researchers in the field of qualitative analysis: (i) Researchers who are pre-computers, who prefer coloured pencils, articles, and note cards; (ii) Researchers who use non-specific software, such as word processors, spreadsheet calculations and general databases, and (iii) Researchers who use specific software to analyse qualitative data, such as NVivo, Atlas.it, webQDA, MaxQDA, etc.

We can summarize the story of specific qualitative research software with some chronological aspects:

(1) In 1966, MIT developed "The General Inquiry" software to help text analysis, but some authors [8, 9] refer that this was not exactly a qualitative analysis software.

(2) In 1984, the software ETHNOGRAPH comes to light and still exists in its sixth version (http://www.qualisresearch.com/).

(3) In 1987, Richards & Richards developed the Non-numerical Unstructured Data Indexing, Searching and Theorizing (NUD*IST), the software that evolved into the current NVivo indexing system.

(4) In 1991, the prototype of the conceptual network ATLAS-ti is launched, mainly related to Grounded Theory.

(5) Approximately in the transition of the 2000 decade, it was possible to integrate video, audio and image in text analysis of qualitative research software. Nevertheless, HyperRESEARCH had been presented before as software that also allowed to encode and recover text, audio and video. Transcriber and Transana are other software systems that emerged to handle this type of data.

(6) In 2004, "NVivo summarizes some of the most outstanding hallmark previous software, such as ATLAS-ti—recovers resource coding in vivo, and ETHNOGRAPH—a visual presentation coding system" [9].

(7) 2009 marks the beginning of the development of qualitative software in cloud computer contexts. Examples of this are Dedoose and webQDA, that were developed almost simultaneously in USA and Portugal, respectively.

(8) From 2013 onwards we saw an effort from software companies to develop iOS versions, incorporating data from social networks, multimedia and other visual elements in the analysis process.

Naturally, this story is not complete. We can include other details and software such as MaxQDA, AQUAD, QDA Miner, etc. For example, we can see a more exhaustive list in Wikipedia's entry "Computer-assisted qualitative data analysis software".

What is the implication of Qualitative Data Analysis Software on scientific research in general and, more specifically, in qualitative research? Just as the invention of the piano allowed composers to begin writing new songs, the software for qualitative analysis also affected researchers in the way they dealt with their data. These technological tools do not replace the analytical competence of researchers, but can improve established processes and suggest new ways to reach the most important issue in research: to find answers to research questions. Some authors [10, 11] recognize that QDAS allow making data visible in ways not possible with manual methods or non-specific software, allowing for new insights and reflections on a study or corpus of data.

Kaefer et al. [10] wrote a paper with step-by-step QDAS software descriptions about the 230 journal articles they analyzed on climate change and carbon emissions. They concluded that while qualitative data analysis software does not do the analysis for the researcher, "it can make the analytical process more flexible, transparent and ultimately more trustworthy" [10, p. 1]. There are obvious advantages in the integration of QDAS in standard analytical processes, as these tools open up new possibilities, such as agreed by Richards [6]: (i) computers have enabled new qualitative techniques that were previously unavailable; (ii) computation has produced some influence on qualitative techniques.

We, therefore, can summarize some advantage of QDAS: (i) faster and more efficient data management; (ii) increased possibility to handle large volumes of data; (iii) contextualization of complexity; (iv) technical and methodological rigor and systematization; (v) consistency; (vi) analytical transparency; (vii) increased possibility of collaborative teamwork, etc. However, many critical problems present challenges to the researchers in this area.

There are many challenges in the field of QDAS. Some are technical or computational issues, whereas other are methodological or epistemological prerequisites, although [6, 12] recognizing that methodological innovations are rarely discussed. For example, Corti and Gregory [13] discuss the problem of exchangeability and portability of current software. They argue the need for data sharing, archiving and open data exchange standards among QDAS, to guarantee the sustainability of data collections, coding, and annotations on these data. Several researchers place expectations on the QDAS utilities on an unrealistic basis, while others think that the system has insufficient analytical flexibility. Richards (2002) refers that many novice researchers develop a so-called "coding fetishism" that transforms coding processes into an end in itself. For this reason, some believe that QDAS can reduce critical reading and reflection. For many researchers, the high financial cost of the more popular QDAS is a problem, but in this article, we would like to focus on the challenge of the considerable time and effort required to learn them.

Choosing a QDAS is a first difficulty that, in several cases, is coincidental with the process of qualitative research learning. Kaefer, Roper and Sinha [10] suggest to compare and test software through sample projects and literature review. Today, software companies offer many tutorial videos and trial times to test their systems. Some authors [14] studied the determinant factors in the adoption and recommendation of qualitative research software. They analyzed five factors: (i) Difficulty of use; (ii) Learning difficulty; (iii) Relationship between quality and price; (iv) Contribution to research; and (v) Functionality. These authors indicate the two first factors as the most cited ones in the corpus of the data analyzed:

- "NVivo is not exactly friendly. I took a whole course to learn to use it, and if you don't use it often enough, you're back to square one, as those "how-to" memories tend to fade quickly." **Difficulty of use**
- "I use Nvivo9 and agree that it is more user-friendly than earlier versions. I do not make full use of everything you can do with it however—and I've never come across anyone who does" "NVivo". **Learning difficulty**

In this context it is very important to study the User Experience and Usability of the QDAS, because these types of tools need to be at the service of the researcher, and not the opposite, therefore reducing the initial learning time and increasing the effectiveness and efficiency of the whole processes of analysis. Usability is an important dimension in the design and development of software. It is important to understand when to involve the user in the process.

3 Design Through Research: Usability Evaluation

Scientific research based on Research and Development (R&D) methodology has a prevalence on quantitative studies. Primarily, it is in the interest of the researcher to test/prove some theory through the actions of individuals involved in the study. The researcher intends to generalize and usually uses numerical data. When we apply R&D to the design of software packages, it becomes poor to reduce the researcher to someone who does not attempt to perceive the context in which the study takes place, interpreting the meanings of the participants, resorting to interactive and iterative processes. Thus, the R&D methodology (or Design Research, also emerging under the expression Research through Design) has been gaining ground in software projects that, according to Pierce [15], "include devices and systems that are technically and practically capable of being deployed in the field to study participants or end users" [p. 735]. Zimmerman et al. [16] state that Design Research "mean[s] an intention to produce knowledge and not the work to more immediately inform the development of a commercial product" [p. 494].

Associated with this methodology, expressions such as Human-Computer Interaction (HCI), User-Centred Design (UCD) and Human-Centred Design (HCD) crop up. Maguire [17], in his "Methods to support human-centred design" study tackles the importance of software packages being usable and how we can reach that wish. This study lists a series of methods that can be applied in the planning, comprehension of the context of use, definition of requirements, and development and evaluation of project solutions.

According to Van Velsen et al. [18], depending on the phase of the project, evaluation can serve different purposes. At the initial phase, when there still does not exist any software, evaluation provides information that supports decision making; at an intermediate phase and through the use of prototypes, it allows the detection of problems; at a final phase, and already in possession of a complete version of the software, it allows to assess quality. The process offered by these authors, named Iterative Design, is divided into four phases [18]:

- Without Software: one aims at making decisions through a collection of data with questionnaires, interviews and focus groups and its analysis. These instruments are developed to characterize and define the user requirements;
- Low-Fidelity Prototype: one aims at detecting problems through gathering and analysis of data obtained from interviews and focus groups. One identifies the need to feel appreciation, perceived usefulness, security and safety aspects;
- High-Fidelity Prototype: one aims at detecting problems through gathering and analysis of data obtained from questionnaires, interviews, think-aloud protocols and observation. Apart from the previous metrics, one aims at assessing comprehensibility, usability, adequacy, and the behavior and performance of the user;
- The Final Version of the Software: in the final version, the same data gathering instruments are applied. Besides some of the metrics mentioned before, one aims at perceiving the Experience and Satisfaction of the User.

According to Godoy [19], "a phenomenon can be better understood in the context in which it occurs and of which it is a part, having to be analyzed from an integrative perspective" [p. 21].[1] It is this framework that in specific moments in the Iterative Design phases the researcher goes to the field to "capture" the phenomenon under study from the point of view of the users involved, considering every relevant point of view. The researcher collects and analyzes various types of data to understand the dynamics of the phenomenon [19].

The recent need to identify and understand non-measurable/non-quantifiable aspects of the experience of the user is leading several researchers in the area of HCI to take hand of quantitative methods. HCI researchers started to understand that the context, both physical and social, in which specific actions take place along with associated behaviors, allow, through a structure of categories and their respective interpretation, to analyze a non-replicable phenomenon, not transferable nor applicable to other contexts [20].

The criteria defined by the standards are essentially oriented to technical issues. However, for a qualitative analysis to support that a software is of quality, it is

---

[1]Our translation.

necessary to take into account the research methodologies to be applied. Being an authoring software, researchers/users need to have knowledge of the techniques, processes, and tools available in terms of data analysis in qualitative research.

4 webQDA Usability Evaluation: Methodological Aspect

Despite ISO 9126 providing 6 dimensions, in this study we will focus on the proposal to evaluate the usability of webQDA qualitative analysis software.

For an effective understanding of usability, there are quality factors that can be assessed through the evaluation criteria. Collecting and analyzing data to answer the following questions will help determine if the software is usable or not [21, 25]:

- Is the theme of the software easy to understand? (Understandable)
- Is it easy to learn how to use it? (Ease of learning)
- What is the speed of execution? (Use efficiency)
- Does the user show evidence of comfort and positive attitudes towards its use? (Subjective satisfaction).

The answers to these questions can be carried out through a quantitative, qualitative [22] or mixed methodology. However, we believe that they all have strengths and weaknesses that can be overcome when properly articulated.

Our study is based on the application of a review of the use of webQDA, applied only to users of this software, by means of a questionnaire to users of webQDA version 2.0 (93 replies) [23] and users of webQDA version 3.0 (39 replies). For the analysis of data we withdrew the answers of the users who answered: "I had/have access to a license but never used it". Due to the fact that SUS measures the usability of a system, the answers of these users would introduce bias in the final results. Therefore, 83 answers were analyzed from version 2.0 users and 34 answers from the users of the present version (3.0).

In terms of this article, for the evaluation of usability, we applied a questionnaire to users of versions 2.0 and 3.0 of webQDA. The results will support decision-making by the development team in the design of version 3.0 of webQDA (http://www.webqda.net) and on the changes that will have to be introduced. To evaluate its usability, we used the System Usability Scale (SUS) [24].

(a) Using SUS

SUS is generally used after the respondent has had an opportunity to use the system being evaluated, but before any debriefing or discussion takes place. Respondents should be asked to record their immediate response to each item, rather than thinking about items for a long time. All items should be checked. If respondents feel that they cannot respond to a particular item, they should mark the center point of the scale.

(b) Scoring SUS

SUS yields a single number representing a composite measure of the overall usability of the system being studied. Note that scores for individual items are not meaningful on their own. To calculate the SUS score, first add the score contributions from each item. Each item's score contribution will range from 0 to 4. For items 1, 3, 5, 7 and 9, the score contribution is the scale position minus 1. For items 2, 4, 6, 8 and 10, the contribution is the scale position minus 5. Multiply the total result of the scores by 2.5 to obtain the overall value of the SUS [24]. If the result is less than 68 points, it should lead the team to conclude that the software faces usability problems. Less than 50 points may be an indicator that you need to invest in the interface design and its usability.

SUS consists of only 10 closed questions and their respective Likert scale representation (1—Strongly disagree, 5—Strongly agree):

(1) I think that I would like to use this system frequently.
(2) I found the system unnecessarily complex.
(3) I thought the system was easy to use.
(4) I think that I would need the support of a technical person to be able to use this system.
(5) I found the various functions in this system were well integrated.
(6) I thought there was too much inconsistency in this system.
(7) I would imagine that most people would learn to use this system very quickly.
(8) I found the system very cumbersome to use.
(9) I felt very confident using the system.
(10) I needed to learn a lot of things before I could get going with this system.

5 Results

The questionnaire was made available online and sent to the users' database with ongoing projects in webQDA (versions 2.0 and 3.0). Focusing our analysis on software version 2.0 of webQDA, we analyzed the qualitative data that allowed supplementing what was considered in the evaluation of its usability. When we used the System Usability Scale (SUS) [24] we got the average central tendency of 72.5 points (SD = 13.2), which allows us to conclude that webQDA 2.0 was "acceptable" in terms of usability, according to the SUS criteria [23], and we got the average central tendency of 66.3 points (SD = 14.1), which allows us to conclude that webQDA 3.0 has "usability problems". Figure 1 shows the scatter plot of these values with a minimum of 32.5 points and a maximum of 100 points.

Figures 2 and 3 show the results obtained in Fig. 1 regarding the level of experience/skill attributed to the use of the qualitative analysis support software. In the application of the survey questionnaire, 4 levels of experience were assigned:

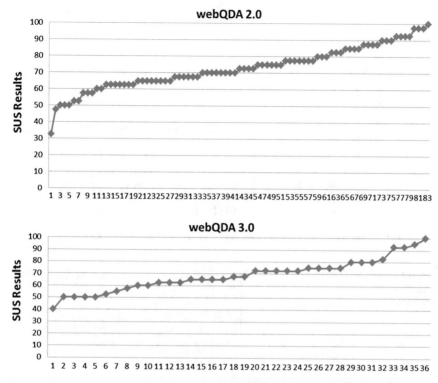

Fig. 1 webQDA 2.0 and 3.0 line graph (SUS results vs. No. of participants)

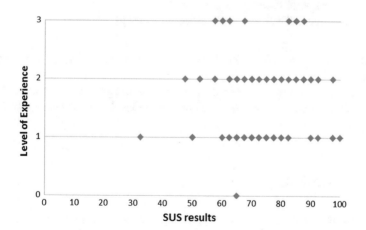

Fig. 2 Dispersion graph as to level of experience versus SUS results (webQDA 2.0)

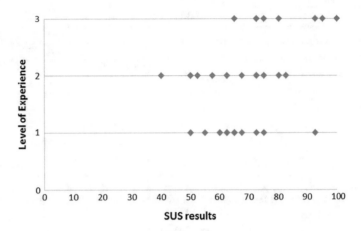

Fig. 3 Dispersion graph as to level of experience versus SUS results (webQDA 3.0)

Table 1 Usability issues in relation to the level of experience

Level	webQDA 2.0 (%)	webQDA 3.0 (%)
0	1 participant (1.2)	0 participants (0)
1	12 participants (14.3)	10 participants (29.4)
2	8 participants (9.5)	6 participants (17.6)
3	1 participant (1.2)	1 participant (2.9)

- Level 0—I do not know how to use it
- Level 1—Just need to learn to use a few more features
- Level 2—Enough to use in my research/survey
- Level 3—Specialist (I can give advice or training)

It is of interest to assess if the level of user-assigned experience influenced the outcome of SUS.

Analysing SUS results lower than 68 points, a figure which shows that the software can/has usability issues in relation to the level of experience, we obtained (Table 1).

6 Final Remarks

In conclusion and in response to our guiding question, we note that the use of the System Usability Scale (SUS) in the evaluation of versions 2.0 and 3.0 of webQDA enabled the multidisciplinary team to be in possession of a deeper grasp of the new interface being developed for release 3.0 of webQDA. It is noteworthy that SUS was effective for its speed in obtaining data. They will be enhanced and complemented with the analysis of the open questions also collected in the questionnaire, and we hope to identify the most critical points. The development team is preparing to triangulate these findings with others that have been collected, for example the operating life of webQDA so far, and its use in a project, as well as the degree of methodological knowledge (types and research designs and contexts of study) of researchers when using the software, with special emphasis on research questions, objectives and type of data being scrutinised.

Acknowledgements The first author thanks the Foundation for Science and Technology (FCT) for the financial support that enabled the development of this study (under the project UID/CED/00194/2013). The authors thank the Micro I0 company and its employees for the development of the new version of webQDA and all the participants of this study.

References

1. ISO9241-210: Ergonomics of Human-System Interaction (210: Human-Centred Design for Interactive Systems) (2010)
2. Scanlon, E., Mcandrew, P., Shea, T.O.: Designing for educational technology to enhance the experience of learners in distance education: how open educational resources, learning design and Moocs are influencing learning. J. Interact. Media Educ. **1**, 1–9 (2015)
3. Souza, F.N. de, Costa, A.P., Moreira, A.: webQDA. http://www.webqda.net (2016)
4. Souza, F.N., Costa, A.P., Moreira, A.: Análise de Dados Qualitativos Suportada pelo Software webQDA. VII Conferência Internacional de TIC na Educação: perspetivas de Inovação, pp. 49–56. VII Conferência Internacional de TIC na Educação, Braga (2011)
5. Teixeira, R.A.G., Neri de Souza, F., Vieira, R.M.: Docentes Investigadores de Programas de Pós-graduação em Educação no Brasil: estudo Sobre o Uso de Recursos Informaticos no Processo de Pesquisa. Rev. da Avaliação da Educ. Super. No prelo, pp. 741–768 (2015)
6. Richards, L.: Rigorous, rapid, reliable and qualitative? Computing in qualitative method. Am. J. Heal. Behav. **26**, 425–430 (2002)
7. Puebla, C.A.C., Davidson, J.: Qualitative computing and qualitative research: addressing the challenges of technology and globalization. Hist. Soc. Res. **37**, 237–248 (2012)
8. Tesch, R.: Introduction. Qual. Sociol. **14**, 225–243 (1991)
9. Cisneros Puebla, C.A.: Analisis cualitativo asistido por computadora. Sociologias 288–313 (2003)
10. Kaefer, F., Roper, J., Sinha, P.: A software-assisted qualitative content analysis of news articles: example and reflections. Forum Qual. Sozialforsch. **16** (2015)
11. Souza, F.N., de Souza, D.N., Costa, A.P.: Asking questions in the qualitative research context. Qual. Rep. **21**, 6–18 (2016)
12. Richards, L.: Qualitative computing—a methods revolution? Int. J. Soc. Res. Methodol. **5**, 263–276 (2002)
13. Corti, L., Gregory, A.: CAQDAS comparability. What about CAQDAS data exchange? Forum Qual. Soc. Res. **12**, 1–17 (2011)
14. Pinho, I., Rodrigues, E., Neri de Souza, F., Lopes, G.: Determinantes na Adoção e Recomendação de Software de Investigação Qualitativa: estudo Exploratório. Internet Latent Corpus J. **4**, 91–102 (2014)
15. Pierce, J.: On the presentation and production of design research artifacts in HCI. In: Proceedings of the 2014 Conference on Designing Interactive Systems—DIS '14, pp. 735–744. ACM Press, New York, USA (2014)

16. Zimmerman, J., Forlizzi, J., Evenson, S.: Research through design as a method for interaction design research in HCI. In: Proceedings of the SIGCHI Conference on Human Factors in Computing Systems—CHI '07, p. 493. ACM Press, New York, USA (2007)
17. Maguire, M.: Methods to support human-centred design. Int. J. Hum Comput Stud. **55**, 587–634 (2001)
18. Van Velsen, L., Van Der Geest, T., Klaassen, R., Steehouder, M.: User-centered evaluation of adaptive and adaptable systems: a literature review. Knowl. Eng. Rev. **23**, 261–281 (2008)
19. Godoy, A.S.: Pesquisa qualitativa: tipos fundamentais. Rev. Adm. Empres. **35**, 20–29 (1995)
20. Costa, A.P., Faria, B.M., Reis, L.P.: Investigação através do Desenvolvimento: Quando as Palavras "Contam." RISTI - Rev. Ibérica Sist. e Tecnol. Informação. vii–x (2015)
21. Seffah, A., Mohamed, T., Habieb-Mammar, H., Abran, A.: Reconciling usability and interactive system architecture using patterns. J. Syst. Softw. **81**, 1845–1852 (2008)
22. Costa, A.P., de Sousa, F.N., Moreira, A., de Souza, D.N.: Research through design: qualitative analysis to evaluate the usability. In: Costa, A.P., Reis, L.P., Sousa, F.N., de Moreira, A., Lamas, D. (eds.) Springer, pp. 1–12. Springer (2016)
23. Costa, A.P., Souza, F.N. de, Moreira, A., Souza, D.N. de: webQDA—qualitative data analysis software usability assessment. In: Rocha, Á., Reis, L.P., Cota, M.P., Suárez, O.S., and Gonçalves, R. (eds.) Actas de la 11 a Conferencia Ibérica de Sistemas y Tecnologías de Información, pp. 882–887. AISTI – Associação Ibérica de Sistemas e Tecnologias de Informação., Gran Canária - Espanha (2016)
24. Brooke, J.: SUS—A Quick and Dirty Usability Scale (1996)
25. Rocha, Á.: Framework for a global quality evaluation of a website. Online Inf. Rev. **36**(3), 374–382 (2012)

Location Privacy Concerns in Mobile Applications

Luis Marcelino and Catarina Silva

Abstract Privacy is a ubiquitous topic in research nowadays, specially in mobile application scenarios. Users are seldom aware of the privacy hazards they are under when using mobile applications. In this paper we compare two privacy studies and present development strategies that can be employed to favour the awareness of privacy and security issues with a strong focus on location data. Results show that developers should play an increasingly important role in the assurance of privacy since users of mobile apps, although concerned with privacy issues, are not fully prepared for making the right decisions in every situation as a rerun of an old study shows.

Keywords Privacy · Security · Mobile applications

1 Introduction

Nowadays, users are greatly dependent on smartphones, at a point that sometimes they neglect privacy concerns, namely how much information is available to others when accepting permissions, privacy policies etc. This can become a problem considering that in the near future, by 2020, there will be nearly 50 billion devices connected to internet [1]. In such a mobile world, this means users may have to

L. Marcelino (✉) · C. Silva
School of Technology and Management, Polytechnic Institute of Leiria,
Leiria, Portugal
e-mail: luis.marcelino@ipleiria.pt

C. Silva
e-mail: catarina@ipleiria.pt

L. Marcelino
Instituto de Telecomunicações, Lisbon, Portugal

C. Silva
Center for Informatics and Systems, University of Coimbra, Coimbra, Portugal

© Springer International Publishing AG 2018
Á. Rocha and L.P. Reis (eds.), *Developments and Advances in Intelligent Systems and Applications*, Studies in Computational Intelligence 718,
DOI 10.1007/978-3-319-58965-7_17

accept application permissions not only on their smartphones, but also on smart TVs, smart watches, and smart glasses, among others.

Sometimes, such permissions are not fully required for the mobile application to provide the desired services, but serves to create an advertising environment that adapts to the location and user's interests. Users take several risks when they systematically accept these permissions. They may be exposing their location and/or their internet habits or the final destination for unclear purposes.

Consequently, previous research has shown that smartphone users are often unware of the data collected by their apps and express surprise and discomfort when they realize the type and volume of such data [2]. Different operating systems have offered specific controls that give users some control on privacy settings. However, privacy decision can be subject to cognitive and behavioural biases, and decision heuristics that often lead to privacy-adverse decisions in favour of short-term benefits [3].

Location privacy has been a hot topic in research and media over the last decade [4], magnified by the popularity of location-aware devices that led to the pervasiveness of applications and services that access user's location to provide them personalized/customized services. Due to recent advances in positioning technology, more and more mobile devices are equipped with GPS (Global Positioning System) and have positioning capabilities. Hence, location based services (LBS) have been widely regarded as one of the most promising services for mobile users [5]. Given the services LBS provide, users are compelled to use them, sometimes posing serious threats to user location privacy by revealing their exact locations. A malicious user or attack, including the LBS server, is then likely to infer the identity of the user [5].

In this paper we compare two privacy studies, one from 1999 and other from 2016 to see how privacy concerns have evolved. We then explore the use of different geo-location precision levels and how can developers ensure privacy for their users in current challenging scenarios.

The rest of the paper is organized as follows. In the next section we will present related work including different strategies to deal with privacy issues in the field of mobile application's permissions as well as geo-localization. In Sect. 3 we present the initial survey on Users' Concerns Online that underpins this work. In Sect. 4 we will detail the proposed approach on Users' Concerns on Mobile Devices, namely the definition of a mobile app to survey user awareness on user privacy and compare approaches. Finally, we will discuss the results obtained and present final conclusions.

2   Related Work

Research in location privacy in mobile environments has been increasing with the related challenges. Different approaches can be found in [6–12]. Since the mobile application market drastically increased, there have been many problems related to malicious applications that leak user's private information stored in smart devices [12]. For instance, in 2012 it was revealed that the Path application automatically uploaded smartphone users' entire address books to its servers. This resulted in generalized protest from users and the provider reviewed the privacy permission model [13]. Other prior work has found that many Android and iOS applications share the user's location with third parties and expose the device identifier to trackers [14]. This behaviour is adopted by applications that are not considered malware, and it occurs in spite of the existing permission architecture. Since 2010 the growth of the Android platform has unfortunately triggered the interest of unscrupulous application developers. Android grayware collects excessive amounts of personal information (e.g., to be used for aggressive marketing campaigns), and malware harvests data or sends premium SMS messages for profit. Grayware and malware have both been found in the Android Market, and the rate of new malware is increasing over time [15]. In a survey [16] results suggest that it may be possible to provide privacy control interfaces with simple explanations to empower users to make an informed choice about obfuscation based on their own privacy concerns. The results in this paper indicate that collection and secondary use are the main factors affecting perceived risk, whereas errors are the main factor affecting trust. Trust affects perceived risk, and both factors determine usage intention [17]. Since we are becoming surrounded by applications on different devices, users are increasingly interested in the quality of technology products. With a good product, a user is also concerned with the product's strategy towards privacy. Application privacy research has produced many useful tools to analyse the privacy-related behaviours of mobile apps. However, these automated tools cannot assess people's perceptions of whether a given action is legitimate, or how that action makes them feel with respect to privacy. Application permission is different between iOS and Android. In the older versions of Android users had to accept all the permissions in order to install application. If any of the permissions could not be accepted, it was not possible to install the application. Users of iOS devices have always had possibility to install application and only when a permission was required the system prompted the user for authorization. Users could then accept that a permission or reject it. From Android OS 6.0 —Marshmallow, users have also this option and more, they can choose which permission they never want to apply with the application.

3   Users' Concerns Online

Privacy concerns are not new. In fact, one of the most renown surveys dates back to 1999 by AT&T Labs-Research [18]. At that time challenges were obviously associated with Internet users rather that inexistent mobile scenarios. Nevertheless, important issues remain up-to-date. To better understand the nature of online privacy concerns, this survey looks beyond the fact that people are concerned and attempts to understand what aspects of the problem they are most concerned about. The authors were concerned with different privacy issues:

- They wanted to know how people would respond to situations where personal information was collected, determining that it was important to ask participants about their concerns through specific online scenarios. Therefore, in addition to the closed form survey questions, they also asked for their reasoning through open-ended questions;
- They were interested in testing the design of privacy and in creating better privacy user interfaces. Again, they probed for the reasons behind the respondents' sensitivities through open-ended questions in addition to standard-form survey questions;
- They also wanted to determine participants' general attitudes and demographics and therefore have used questions that had appeared on other surveys so the sample could be matched against others.

In [18] users are divided into 3 categories:

1. **Privacy fundamentalists**: extremely concerned about any use of their data and generally unwilling to provide their data to Web sites, even when privacy protection measures were in place;
2. **Pragmatists**: also concerned about data use, but less so than the fundamentalists. They often had specific concerns and particular tactics for addressing them;
3. **Marginally concerned**: generally willing to provide data to Web sites under almost any condition, although they often expressed a mild general concern about privacy.

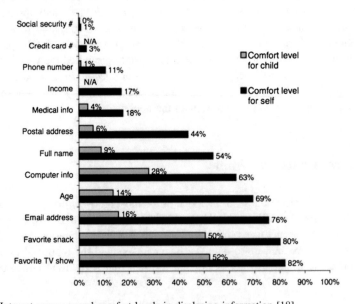

Fig. 1 Internet survey users' comfort levels in disclosing information [18]

As shown in Fig. 1, we can see that comfort levels for insensitive data such as Favourite TV show or Favourite snack have a reasonably high percentage, whereas for sensitive data such as social security and credit card numbers people tend to care much more and to reveal they don't feel as much as safe.

General conclusions of the survey include the deduction that to raise comfort levels in sharing information, users must be surely informed so that they can trust whatever policies are disclosed. Additionally, from a pragmatic perspective, one should focus on those things about which users say they are most concerned. These results provide evidence of what those concerns are among an connection savvy population.

4 Users' Concerns on Mobile Devices

4.1 Introduction

The above mentioned study about privacy [18] is, as referred, from 1999, when the Internet was not widely available, when it was available was through cable and a distant 8 years before the first iPhone was presented to the public. The access to the Internet was so restricted at that time that privacy was not one of the main concerns for those online users. Now that we have access to the Internet from anywhere at any time, one would assume that privacy concerns have changed dramatically.

A small informal study [19] reproduced part of this survey to try to find out how did privacy concerns changed. Therefore, similar questions to those presented in [18] were asked to the participants of this study, but this time focused on mobile devices, given their ubiquity. This privacy study involved 40 participants of which 21 males and 19 females with ages between 18 and 39.

This study showed that only 7.5% of participants were comfortable to provide their phone number to a mobile app, while nearly half of participants (45%) where comfortable to provide their full name. Additionally, participants disclosed age more easily, with a 65% comfort level, and their Favourite TV Show even more easily, with 77.5% of participants stating they were comfortable to provide that information to the mobile app. One curious results from the study is that participants were more willing to share their precise location (15%) than to share their phone number.

Comparing both studies [18] and [19], as depicted in Fig. 2, we note that the percentage of participants comfortable to share their phone number, their full name, their age and their favourite TV shows is approximately the same now and nearly 20 years ago. What participants were willing to disclose to strangers then is roughly what they are willing to disclose a mobile app now. However, participants are less comfortable to provide their exact location now than participants in 1999 were to provide their home address.

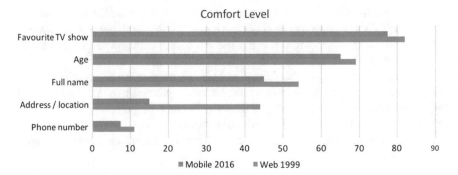

Fig. 2 Comparing comfort levels to disclose personal information

Like the authors of the original study, we are surprised that participants are not very comfortable to provide a mean of contact such as the phone number. They are even less comfortable than identifying exactly where they are. This type of decisions may actually put users of mobile phones at risk.

This risky behaviour from users may put the burden of safety onto the shoulders of app developers. Just because mobile phones can provide the user position accurately, it does not mean that applications should use those values, unless they are relevant to the functionality or to the service it provides. As it is safer to any service provider not to store passwords in plaintext it may be safer note to store the exact positions for their users.

To get the user position, applications can take it from the GPS (if available) or from external systems or networks. The accuracy of these methods varies and developers must decide what is the required precision for the application to work. In the programming environments for iOS and Android there are particular levels of precisions to get the best information for the application and yet to reduce privacy issues, as well as save phones battery.

4.2   User's Location

There are two options for setting up location-related services:

- Use the standard location service, which allows you to specify the desired accuracy of the location data and receive updates as the location changes;
- Use the significant location change service, which provides a more limited set of tracking options but offers significant power savings over the standard location services.

There are several specific functionalities when programming mobile applications for iOS or Android where developer can set the different precisions of the location services. The Location Manager class (CLLocationManager on iOS or

LocationManager on Android) is the central point for configuring the delivery of location and heading-related events. A Location Manager object tracks large or small changes in the user's current location with a configurable degree of accuracy. This configurable degree of accuracy is explained more detailed next, where we present the difference between types of accuracy.

4.2.1 Source of Location Information

There are different ways of getting the users position depending on amount of accuracy application needs to use.

- Network-based: Location can be determined using a network infrastructure, such as cellular towers. The accuracy of network-based techniques varies according to the available cells that are triangulated and existing obstacles;
- Handset-based: In this solution, the mobile device determines its position technique determines the location based on of the handset by putting its location by cell identification, signal strengths of the home and neighbouring cells, which is continuously sent by the carrier;
- Wi-Fi: Typical parameters useful to geo-locate device is getting the Wi-Fi hotspot's the SSID and the MAC address of the access point where the device is currently connected;
- GPS: The best and most precise technique of getting users' location. It uses a network of about 30 satellites orbiting the Earth and each one transmits information about its position and the current time at regular intervals. The GPS receiver triangulates these time stamped signals to determine its position.

4.2.2 Desired Accuracy

Some of generic accuracy ranges are available both in iOS and Android:

(a) Best for Navigation
(b) Hundred Meters
(c) Kilometer
(d) Few Kilometers

4.3 Developers Role in Privacy Concerns

Within making of application developers are assigning one of these values to the 'desired accuracy' property for usage in certain scenario, based on how much accuracy an application needs. Depending on the information about user's location needed developers should use corresponding value and this is where some of the developers misuse this option. If high accuracy is not needed, then developer should specify accuracy Kilometer and not accuracy BestForNavigation. Determining a location with greater accuracy requires more time, more power and less privacy for the end user.

5 Conclusions and Future Work

In this work we analyse location privacy concerns in particular on mobile application scenarios. The identified challenges are mainly related to the unawareness of users of such concerns that have not changed significantly since the introduction of the smartphone. Such conclusions were reached by the analysis and comparison of two privacy studies that showed the comfort level for users to disclose different levels of personal information. That allowed for the presentation of development strategies that can be employed to favour the awareness of privacy and security issues, namely the use of full precision location only when it is mandatory to keep the desired service.

The comparison of the studies, almost 20 years apart, has shown that users concerns have not changed significantly, despite the acceptable initial hypothesis that this would not be the case.

Hence, the main conclusion is that developers should play an increasingly important role in the assurance of privacy, since users of mobile apps, although concerned with privacy issues, are not fully prepared for making the right decisions in every situation, not 20 years ago, and not now.

References

1. Cisco: Internet Of Things will deliver $1.9 trillion boost to supply chain and logistics operations. http://newsroom.cisco.com/press-release-content?articleId=1621819 (2015)
2. Almuhimedi, H., Schlub, F., Sadeh, N., Adjured, I., Acquits, A., Gluck, J., Cranor, L., Agarwal, Y.: Your location has been shared 5,398 times!: A field study on mobile app privacy nudging. In: Proceedings of the 33rd Annual ACM Conference on Human Factors in Computing Systems, pp. 787–796 (2015)
3. Acquisti, A., Grossklags, J.: Privacy and rationality in individual decision making. IEEE Secur. Priv. 3(1), 26–33 (2005)
4. Fawaz, K., Shin, K.: Location privacy protection for smartphone users. In: 2014 ACM SIGSAC Conference on Computer and Communications Security, pp. 239–250 (2014)
5. Suo, H., Liu, Z., Wan, J., Zhou, K.: Security and privacy in mobile cloud computing. In: IEEE 9th International Wireless Communications and Mobile Computing Conference (IWCMC), pp. 655–659 (2013)
6. Wu, C., Huang, C., Huang, J., Hu, C.: On preserving location privacy in mobile environments. In: IEEE International Conference on Pervasive Computing and Communications Workshops (PERCOM Workshops) (2011)
7. Zhong, S., Li, L., Liu, Y., Yang, Y.: Privacy-preserving location-based services for mobile users in wireless networks, (2004)

8. Mitchell, M., Patidar, R., Saini, M., Singh, P., Wang, A.: Mobile usage patterns and privacy implications. In: The Fourth IEEE International Workshop on the Impact of Human Mobility in Pervasive Systems and Applications, pp. 457–462 (2015)
9. Chow, C.-Y., Mokbel, M.F., Aref, W.G.: The new casper: query processing for location services without compromising privacy. In: 32nd International Conference on Very large Data Bases, VLDB, pp. 1–45 (2006)
10. Damiani, M.: Location privacy models in mobile applications: conceptual view and research directions. GeoInformatica, pp. 18–4 (2014)
11. Damiani, M., Galbiati, M.: Handling user-defined private contexts for location privacy in LBS. In: ACM SIGSPATIAL (2012)
12. Kim, S., Cho, J.I., Myeong, H.W., Lee, D.: A study on static analysis model of mobile application for privacy protection, vol. 114. Springer, Dordrecht (2012)
13. Cnet: Path to pay $800,000 to settle privacy issues with FTC. http://www.cnet.com/news/path-to-pay-800000-to-settle-privacy-issues-with-ftc/ (2013)
14. Egele, G., Kruegel, M., Kirda, C., Vigna, E.: PiOS detecting privacy leaks in iOS applications. In: Proceedings of 18th Annual Network Distributed System Security. Symposium. NDSS 2011, p. 11 (2011)
15. Felt, A., Finifter, M., Chin, E., Hanna, S., Wagner, D.: A survey of mobile malware in the wild. In: Proceedings of 1st ACM Work Security and Privacy in Smartphones and Mobile Devices—SPSM '11, pp. 3–14 (2011)
16. Brush, A., Krumm, J., Scott, J.: Exploring end user preferences for location obfuscation, location-based services, and the value of location. http://research.microsoft.com/pubs/135611/ubicomp243-brush.pdf (2010)
17. Zhou, T.: The impact of privacy concern on user adoption of location-based services. Ind. Manag. Data Syst. 111(2). http://www.emeraldinsight.com/doi/pdfplus/10.1108/02635571111115146 (2011)
18. Cranor, L.F., Reagle, J., Ackerman, M.S.: Beyond concern: understanding net users' attitudes about online privacy, AT&T Labs-Research Technical Report TR 99.4.3, 14 Apr 1999
19. Gašparović, M., Nicolau, P., Marques, A., Silva, C., Marcelino, L.: On Privacy in User Tracking Mobile Applications", CISTI'2016—11ª Conferencia Ibérica de Sistemas y Tecnologías de Información. Gran-Canaria, Spain (2016)

Towards a New Approach of Learning: Learn by Thinking Extending the Paradigm Through Cognitive Learning and Artificial Intelligence Methods to Improve Special Education Needs

Jorge Pires, Manuel Pérez Cota, Álvaro Rocha and Ramiro Gonçalves

Abstract Cognitive theorists believe that the learning process involves the integration of events into actives organizational structures termed schemata. Schemata serve a number of functions in human cognition: schemata regulates attention, organizes searches of the environment and "fill the gaps" during information processing. Thus, the mind uses schemata to selectively organize and processes all the information individuals receive from the world. This perspective fits e.g. in teaching blind and deaf people alongside of children with special education needs. The aim of the research developed until the moment was to prove that the full integration of the concept of teaching and learning in the light of cognitive theories.

Keywords Adaptive cognitive learning · Hypermedia system · Genetic algorithms · Chi-square · eLearning · Java · XML · Knowledge-Block · Ubiquitous computing · Pervasive computing

J. Pires (✉) · M.P. Cota
Escuela Superior de Ingeniería Informática, Universidad de Vigo, Vigo, Spain
e-mail: jorgepires.email@gmail.com

M.P. Cota
e-mail: mpcota@uvigo.es

Á. Rocha
Departamento de Engenharia Informática, Universidade de Coimbra,
Coimbra, Portugal
e-mail: amrocha@dei.uc.pt

R. Gonçalves
Universidade de Trá-os-Montes e Alto Douro, Vila Real, Portugal
e-mail: ramiro@utad.pt

© Springer International Publishing AG 2018
Á. Rocha and L.P. Reis (eds.), *Developments and Advances in Intelligent Systems and Applications*, Studies in Computational Intelligence 718,
DOI 10.1007/978-3-319-58965-7_18

1 Introduction

The use of computers and the Internet have deep impact in the classic methods of teaching and learning [1, 2]: e.g. introducing the concept of distance learning as a great opportunity for studying unfettered by constraints of time and space [3, 4]. In addition, acquisition of new skills and knowledge is not only affected by an individual's mental schemes or beliefs, but also by their interaction, cooperation and collaboration with others [3]. The fast development of information and communication technologies (ICTs) has been improving human learning. Mark Weiser envisioned a world in which computers and computing is pervasive and seamlessly embedded in everyday environments and devices. Recent researches have shown that ubiquitous computing (UC), or pervasive computing (PC) may promise to support human learning in a real and smoother context [5].

The context-aware feature of ubiquitous computer environments allows applications to better understand the user's behaviors in the real world. Cheng et al. demonstrated how a u-learning system provides services [6]:

- Set the instructional requirements for each of the learner's learning actions
- Detect the learner's behaviors
- Comparing the requirements with the corresponding learning behaviors
- Provides personal support to the learner

    uLearning characteristics environment are given as follows:

- A ubiquitous learning environment (ULE) is context-aware; that is, learner's situation(s) of the real world environment where he or she is located can be sensed
- ULE actively provides personalized support in the right way, place and time based on personal and environmental situation of the learner in the real world as well as the profile and learning portfolio of the learner
- ULEs enable seamless learning anywhere or anytime; that is, learning occurs without interruption while learner moves from place to place
- ULE adapts the subject contents to meet the functions of various mobile devices

    Cognition has grown as a doctrine based on the various time-empirical observations, and hard data of field research evidence proof, that are mental structures underlying not only thought, but also emotion, as well as the very perception and interpretation of both, inner/internal and outer/external source information [4, 7]. Whose multiple intelligence theory is based on cognitivism, asserts that mind consists of numerous fairly specific and independent computational mechanisms, and it is in this context that research on learning styles has also been promoted. Based on cognitive learning theory, the structure of content of the cognitive matter should be organized hierarchically [8]. Relevant research [7] has surely led to the conclusion that students learn mainly from the progressive and relation-linked construction of knowledge. This approach may well find applications in a Learning Management System (LMS) with a psycho-pedagogically driven learning path creation module [7].

From constructivist point of view the knowledge "built" by an individual and not broadcasted, is itself both a reflective and active process. The interpretation that the individual performs of the new experience is influenced by their prior knowledge introducing in social interaction, multiple perspectives of learning. Learning requires understanding of the whole and the parts, and should be understood in a global context. In this perspective, [9] introduces one new dynamic, co-constructivism. The theory of Structural Cognitive Modifiability (SCM) [9], far transcends the purely cognitivist approach, and advocates that every individual is modifiable, a process that is inherent to the human species.

The authors propose a new architecture based on an intelligent structure supported by Chi-square and a Genetic Algorithm—the evaluation block—supported in a special structured Learning Object, named Knowledge Block—the information and knowledge block supported by the most recent cognitive theories [10].

2 Literature Review

Cognitive Learning theories have a unique place in history: they explore the depths of the mind from the perspective of process. The most dominant aspects of cognitive theory involves the interaction between mental components and the information that is processed through the complex network that is the human brain [11].

Cognitive Learning Theory (Fig. 1) implies that the different processes concerning learning can be explained by analyzing the mental processes first. It posits that with effective cognitive processes, learning is easier and new information can be stored in the memory for a long time. On the other hand, ineffective cognitive processes result to learning difficulties that can be seen anytime during the lifetime of an individual [12].

We cannot see "good thinking" its content or its processes [11]. Cognitive psychologists use metaphors to describe what they cannot see or touch, this metaphors are ways of talking about things too abstract to describe literally or precisely. When the expression "Life is like a box of candy's" is used, we are using a metaphor to describe the intangible. Cognitive scientists use a variety of metaphors to describe the content and processes of "good thinking"—they say, "Since you can't see or

Fig. 1 Social cognitive theory

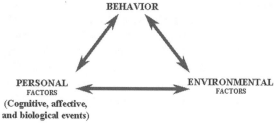

Social Cognitive Theory Illustration (Pajares, 2002)

touch good thinking, why not think of it as a computer, or a filing system, or as an information management system" [11]. If thinking using metaphors is helpful, then the metaphor has value. Almost all cognitive approaches to learning are concerned with how everyday experiences are transformed or processed into mental images or sounds and stored for later use—in other words—the concer is how information is processed. It is logical, therefore, that cognitive psychologists have chosen the information processing model or computer as their metaphor of choice [11].

Moving from traditional learning to educational eSystems motivate learners to get increasingly involved in their learning process. The use of IT in education covers a wide range of very different activities: e.g. learning environments, course management, and much more. Because the one-size-fits-all paradigm cannot be applied to individual learning, adaptability is becoming a must. Hence, courseware is meant to be tailored according to the learner's needs.

Two main families of computerized applications aspire to offer this adaptability: Intelligent Tutoring Systems (ITS) [13] and Adaptive Hypermedia Systems (AHS) [13]. ITS [13] rely on curriculum sequencing mechanisms to provide the student with a path through the learning material. An adaptability algorithm computes this so-called personalized path corresponding to the course construction, the curriculum sequencing [13].

The process is twofold:

- Find the relevant topics and select the most satisfactory one
- Construct dynamically page contents based on the tutor decision for what the learner should study next

ITS usually provide an evaluation of the learners level of mastery of the domain concepts through an answer analysis and error feedback process that eventually allows the system to update the user's model. This process is called intelligent solution analysis [13]. AH [13] was born as a trial to combine ITS and AH. As in ITS, adaptive educational hypermedia focus on the learner, while at the same time it has been greatly influenced by adaptive navigation support in educational hypermedia [13, 14]. In fact, adaptability implies the integration of a student model in the system in the framework of a curriculum which sequence depends on pedagogical objectives, user's needs and motivation. Hence, the use of adaptive and/or interactive hypermedia systems was proposed as a promising solution [13, 14]. Adaptivity in eLearning is a new research trend which personalizes the educational process through the use of Adaptive Educational Hypermedia Systems (AEHS). These systems try to create individualized courses according to the learner personal characteristics, such as language, learning style, preferences, educational goals and progress. In this way, expect to solve some of the major problems and succeed in achieving a better learning outcome [15]. However, there are still problems in the present architectures. Information comes from different sources, embedded with diverse formats into the form of metadata making it troublesome for the computerized programming to create professional materials [16]. The major problems are [16]:

- Difficulty of learning resource sharing
- Even if all eLearning systems follow the common standard, users still have to visit individual platforms to gain appropriate course materials contents. It is comparatively inconvenient
- High redundancy of learning material
- Due to difficulty of resource-sharing, it is hard for teachers to figure out the redundancy of course materials and therefore results in the waste of resources
- Even worse, the consistency of course content is endangered which might eventually slow down the innovation momentum of course materials
- Deficiency of the course brief
- It is hard to abstract course summary or brief automatically in efficient way. So, most courseware systems only list the course names or the unit titles. Information is insufficient for learners to judge quality of course content before they enroll certain courses.

3   Research Context

By studies carried out for years, it has become evident that predict academic performance is a difficult execution task. Each research carried out, ends with the creation or implementation of a model, which seeks to explain the effect of some of the variables considered. These models generally apply statistical methods for obtaining results. It appears that these models do not include important explanatory variable. For this reason, it is necessary to develop a model which admits all the available data and interpret its meaning in conjunction, to produce the desired result [17].

Adaptability is a key concept in the applicability of a correct tactic in the achievement of the objectives. Being that our approach is directed to the customization of the transmission of knowledge, by a continuous guidance of the student in the acquisition of this knowledge, and by a feedback in "real time" of the evolution of this process, we believe in the end we will get what we call individualized cognitive profile, which will enable us in each area of knowledge "know better" the learning speeds and absorption capacities and extrapolation of this knowledge on the part of the student [15].

According to Moreira [18], the teaching/learning process is composed of four elements—the teacher, the student, the content and the environmental variables (characteristics of the school, home ...)—exerting each one, greater or lesser influence on process depending on the way by which relate in a certain context. Even in social situations that are highly complex, our vision allows the absence of a choice that adversely affects the educational factor, given that the teachers may draw a whole range of educational tools, which do not require that the student have to pay for her education. The aim of this research was to develop a system that allows a full integration of the concept of teaching/learning in the light of theories of Feuerstein [9] and Gardner [8], supported by consensus between the theories of

[19, 20] and [21], currently standing implemented in a kindergarten with the appropriate adaptations to the educational context—in this case in particular, in a special education children program.

With this research we develop a new paradigm, creating a self-regulated adaptive system that allows children's involved in a special education program fill their gaps in her learning process. The use of a genetic algorithms combined with a statistical function and evaluation learning models properly defined, that, permits through the use of this architecture the individualization of pedagogical contents to transmit to each child so they can improve their cognitive profiles—or their multiple intelligences [8, 17]. The approach taken is own and covers a series of concepts inherent in the eLearning environments as are the adaptability, usability, reusability, collaboration, ubiquity and Human Computer Interaction (HCI) among others.

4 Methodology

This section describes the architecture system and our proposed approach based on the evolvement technique through a structure named Knowledge Block (KB). First, an overview of the methodology and system architecture is presented in Sects. 4 and 5. Section 6 describes the strategy used to calculate the appropriate KB. After these two processes we will apply the Genetic Algorithm (GA) and the pre-assessment statistical function to construct an appropriate learning path providing the most suitable knowledge unit to learner.

GA are inspired by the biological process of Darwinian evolution where selection, mutation and crossover play a major role. The good solutions are selected and manipulated to achieve new and possibly better solutions. The manipulation is done by the genetic operators that work on the chromosomes in which the parameters of possible solutions are encoded. In each generation the new solution replaces the old in the population selected for deletion [9].

Briefly the process is as follows:

- Initialization Stage—The search space of all possible solutions are mapped onto a set of finite strings—in our case a 32 bit's set string (Table 2). Each string (a chromosome) has is corresponding point in the search space. The algorithm starts with the initial solutions that are selected from a set of configurations in the search space—the population—using a randomly generated solutions (Fig. 2). Each of the initial solutions—the initial population—is evaluated by a defined fitness function (Fig. 2). The fitness function exists with the objective to summarize the encoded
- Selection Stage—The selection stage of a GA is where the individual genomes from the population are "chosen" for later breeding (i.e. crossover, mutation). In this process a set of individuals that have the higher scores in the fitness function are selected to reproduce itself. This strategy—denominated elitism—is a very successful (slight) variant of the general process of constructing a new

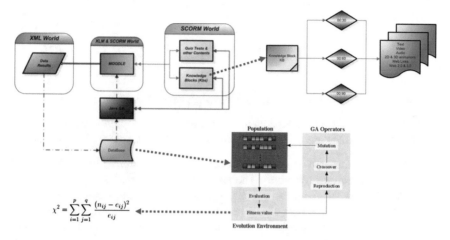

Fig. 2 Genetic algorithm flow

population. This results in the best-performing chromosomes in the population—even with the risk of a small loss in diversity

- Crossover Stage—Crossover operation is a genetic operator that is used to swap corresponding segments of a string representation of a couple of chromosomes from one generation to the next (i.e. single and multi-point crossover, "cut and splice". We use the tournament selection method to select an individual from a population. This method involves running several "tournaments" among a few individuals randomly chosen from the population. Selection pressure is easily adjusted by changing the tournament size
- Mutation Stage—Mutation operates on a single chromosome: one element is randomly chosen from the chains of symbols and the bit string representation is changed with another one. This is an operator used to maintain genetic diversity from one generation of a population to another
- Initial Population Size—The initial population size can be determined according to the complexity of the problem. A larger population size will reduce the search speed of the GA
- Selecting the fitness function—The fitness function is associated to a pre-assessment statistical function (Chi-square), that "judges" the quality of the results, in order to choose the next KB for an learner

The Chi-square is defined as a discrepancy measure between the observed frequencies and the expected ones (Fig. 3) [22].

Where p and q is the number of observed values (Oj), e_{ij}, is the expected frequency of symbols, and n_{ij} is the observed frequency. The independence test of Chi-square allows you to check the independence between two variables of any type, grouped in a contingency table.

Fig. 3 Chi-Square function

$$\chi^2 = \sum_{i=1}^{p}\sum_{j=1}^{q}\frac{(n_{ij} - e_{ij})^2}{e_{ij}}$$

The theory is based mainly on the following two assumptions:

- H_0—the variables are independent—the method is valid and adequate
- H_1—the variables are not independent—the method is not valid or adequate

Note that the alternative hypothesis does not have any information on the type of association between the variables. The test works by comparing the observed frequencies of each of the p × q cells, n_{ij}, with the corresponding frequencies expected under the hypothesis of independence, e_{ij}, through the value which is used for the calculation of the coefficient of contingency of Pearson [22] (Fig. 3). If this value is small enough then the corresponding "barrier" is established by the significance level of the test, which means that the differences n_{ij}–e_{ij} are small, and we must accept H_0 as the valid hypothesis [22].

The methodology used is as follows: A group composed of 17th kindergarten children was divided into two groups. A control group of 9 elements and a test group of 8 elements. In order to make the results "clean" between the researchers and the kindergarten teacher, it was created a table (Table 1) that identifies both groups of children through IDs. The correspondence between the true identity and the corresponding ID are only kindergarten teacher knowledge. The obtained results of 2 years' fieldwork amounted to 13600 samples.

5 System Design and Development

The main structure for knowledge and evaluation is based on KB that is formed in a predetermined orientation (Fig. 4) with the system structure that we can observe in Fig. 5.

In Fig. 3 we can see the architecture of the system from the viewpoint of its functionalities. The system consists of two main modules: the GA based module (Java developed) is composed of a GA the pre-assessment function, and a database to save all the profile history, while the XML & SCORM based module consists of a Learning Management System (LMS)—Moodle and a KB repository.

Table 1 ID correlation table

Real name and identification (e.g.)	Test identifier (identifier structure) (e.g.)
Martim	Gn01
Matilde	Gn02
...	...
Tiago	Gn10

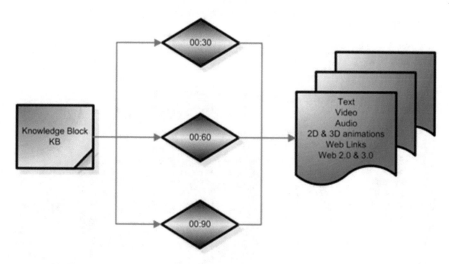

Fig. 4 Orientation layer structure

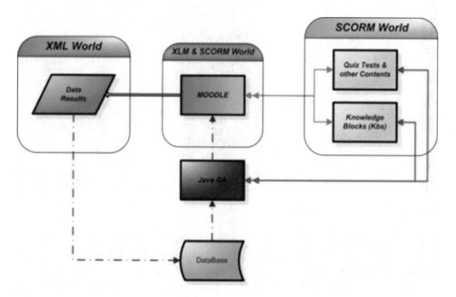

Fig. 5 System architecture

The choice of a KB depends on the objective pursued; may be the improvement of a particular cognitive profile, or a particular field of study (i.e. physics, mathematics).

The KB are simple structures (Fig. 4), binary encoded with a 32-bit string as a descriptor (Table 2). This allows a greater flexibility in identification and future expansion. The grade crossing between KB is determined by the pre-evaluation

Table 2 ID correlation table

Reserved for future use	Educational level	Cognitive profile ID	KB difficulty level
00000000	00000	000	0000000000000000

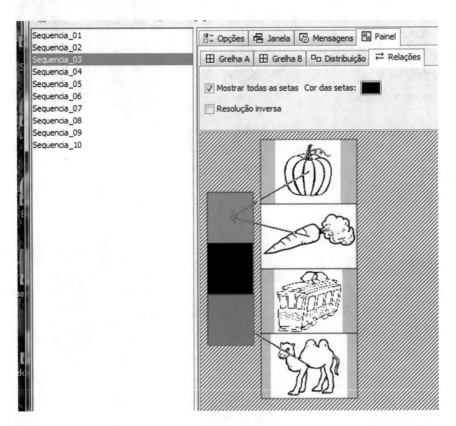

Fig. 6 Quiz module in construction

function—Chi-square (Fig. 2) together with the structure of Fig. 5. In Fig. 5 we can see the architecture of the system from the viewpoint of its functionalities. The system consists of two main modules: the GA based module—that is composed of a GA and a pre-assessment function (Fig. 3) a database to save all the profile history, while the XML & SCORM based modules consists of a Learning Management System (LMS)—Moodle and a KB repository.

The main interface is composed by the LMS where the KB are loaded and displayed. In Fig. 6 we can see a quiz module that is used in conjunction with the GA based module to determine the academic level of the evaluated subject. In this study a binary descriptor is assigned to each curriculum in a 32 bits string (Table 2). This descriptor contains all the information needed to identify the difficulty level, the cognitive ID profile (i.e. logical-mathematical, linguistic …) and the educational level (i.e. 1st grade, high-school, bachelor's grade …).

6 Research Results

6.1 Phase I (Short Vision)

Before assuming the evaluation function as valid there were realized random tests in order to ensure its validity in the context. These tests were not based on any value resulting from any real observation. The focus were to stress the function to the limit to understand if it allow us to validate, or not, their integration in the present research.

With 10 samples (Fig. 7) and with an expected value for (Ej) of 14, the results leads us to conclude that the H_0 hypothesis is valid since the value obtained through the application of the evaluation function is inferior to the value in the distribution table [22]—2,8571 against 3,325 respectively.

Changing the B5 and C5 cell values (EJ) to 15 (Fig. 8) we can observe that the hypothesis H_0 are no longer valid because the obtained value are slightly superior to the observed in the distribution table [22].

Has can be observed (Fig. 9) when (Ej) value change to 15 the obtained value considerably exceeds the observed in the distribution table [22] (Fig. 10).

Maintaining all the structure and changing only the cells B4 and H4 one value each, the H_0 hypothesis is again true. We can then conclude hypothetically that the student can now achieve the level 15.

6.2 Phase II (Short Vision)

In this phase and after the stress tests show that Chi-square function is adequate to be applied as evaluation function, the second stage of the research was to apply all the knowledge and theoretical studies in live environment. The data values are the

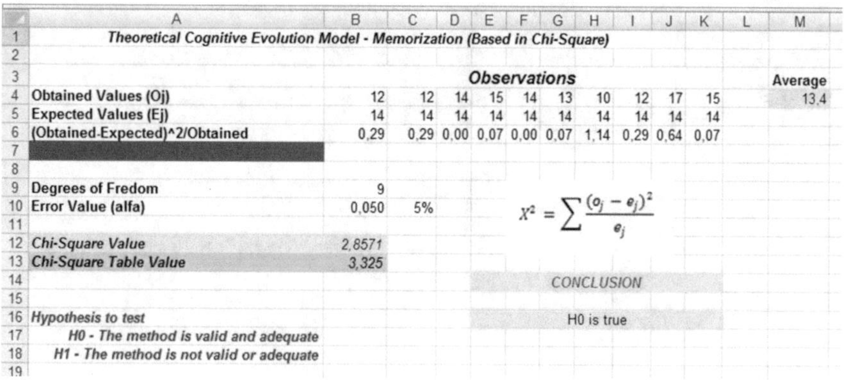

	A	B	C	D	E	F	G	H	I	J	K	L	M
1	Theoretical Cognitive Evolution Model - Memorization (Based in Chi-Square)												
2													
3					Observations								Average
4	Obtained Values (Oj)	12	12	14	15	14	13	10	12	17	15		13,4
5	Expected Values (Ej)	14	14	14	14	14	14	14	14	14	14		
6	(Obtained-Expected)^2/Obtained	0,29	0,29	0,00	0,07	0,00	0,07	1,14	0,29	0,64	0,07		
7													
8													
9	Degrees of Fredom	9											
10	Error Value (alfa)	0,050	5%										
11													
12	Chi-Square Value	2,8571											
13	Chi-Square Table Value	3,325											
14								CONCLUSION					
15													
16	Hypothesis to test							H0 is true					
17	H0 - The method is valid and adequate												
18	H1 - The method is not valid or adequate												
19													

$$x^2 = \sum \frac{(o_j - e_j)^2}{e_j}$$

Fig. 7 Chi-Square matrix test

	A	B	C	D	E	F	G	H	I	J	K	L	M
1	Theoretical Cognitive Evolution Model - Memorization (Based in Chi-Square)												
2													
3					Observations								Average
4	Obtained Values (Oj)	12	12	14	15	14	13	10	12	17	15		13,4
5	Expected Values (Ej)	15	15	14	14	14	14	14	14	14	14		
6	(Obtained-Expected)^2/Obtained	0,60	0,60	0,00	0,07	0,00	0,07	1,14	0,29	0,64	0,07		
7													
8													
9	Degrees of Fredom	9											
10	Error Value (alfa)	0,050	5%		$X^2 = \sum \frac{(o_j - e_j)^2}{e_j}$								
11													
12	Chi-Square Value	3,4857											
13	Chi-Square Table Value	3,325											
14						CONCLUSION							
15													
16	Hypothesis to test					H1 is true and H0 is disposable							
17	H0 - The method is valid and adequate												
18	H1 - The method is not valid or adequate												
19													

Fig. 8 Change expect value (Ej) of all cells to 15

	A	B	C	D	E	F	G	H	I	J	K	L	M
1	Theoretical Cognitive Evolution Model - Memorization (Based in Chi-Square)												
2													
3					Observations								Average
4	Obtained Values (Oj)	12	12	14	15	14	13	10	12	17	15		13,4
5	Expected Values (Ej)	15	15	15	15	15	15	15	15	15	15		
6	(Obtained-Expected)^2/Obtained	0,60	0,60	0,07	0,00	0,07	0,27	1,67	0,60	0,27	0,00		
7													
8													
9	Degrees of Fredom	9											
10	Error Value (alfa)	0,050	5%		$X^2 = \sum \frac{(o_j - e_j)^2}{e_j}$								
11													
12	Chi-Square Value	4,1333											
13	Chi-Square Table Value	3,325											
14						CONCLUSION							
15													
16	Hypothesis to test					H1 is true and H0 is disposable							
17	H0 - The method is valid and adequate												
18	H1 - The method is not valid or adequate												
19													

Fig. 9 Change expect value (Ej) of all cells to 15

	A	B	C	D	E	F	G	H	I	J	K	L	M
1	Theoretical Cognitive Evolution Model - Memorization (Based in Chi-Square)												
2													
3					Observations								Average
4	Obtained Values (Oj)	13	12	14	15	14	13	11	12	17	15		13,6
5	Expected Values (Ej)	15	15	15	15	15	15	15	15	15	15		
6	(Obtained-Expected)^2/Obtained	0,27	0,60	0,07	0,00	0,07	0,27	1,07	0,60	0,27	0,00		
7													
8													
9	Degrees of Fredom	9											
10	Error Value (alfa)	0,050	5%		$X^2 = \sum \frac{(o_j - e_j)^2}{e_j}$								
11													
12	Chi-Square Value	3,2000											
13	Chi-Square Table Value	3,325											
14						CONCLUSION							
15													
16	Hypothesis to test					H0 is true							
17	H0 - The method is valid and adequate												
18	H1 - The method is not valid or adequate												
19													

Fig. 10 Change (Oj) in cell B4 from 12 to 13 and in cell H4 from 10 to 11

result of the field investigation that occurs in the school years of 2012–2014. The application of the (KB) and the quizzes between the two groups—test group and control group was intercalated between weeks. The first week was for the test group and the following week for the control group, and so on. The kindergarten—preschool education, has three main areas:

- Personal and social training—transverse to all areas
- Communication and expression area—field of expressions (colors, plastic expression, dramatic expression, motor expression, opposites and topological notions), mathematics, oral and written language, tales and terminating of words
- World knowledge area—social and physical environment

According Gardner's theory [8] we create the structure as follows:

- Animals (the world and physical environment)—physical environment—naturalist intelligence
- Geometric figures, Lines—logical/mathematical intelligence
- Colors, Patterns, Opposites and Motor expression—visual-spatial intelligence
- Tales and words ending in ão—verbal-linguistic intelligence

The Musical Intelligence was always present in the sounds of the quizzes—in a minimalist way (not measured)—but present. The Bodily-Kinesthetic Intelligence, Interpersonal Intelligence and Intrapersonal Intelligence are not possible to quantify —since that are intelligences that we cannot quantify—are reported by the kindergarten teacher in his personal and independent report.

The results after that field work period are, as show in Figs. 11, 12, 13 and 14 as following. In the left we can see the group that was tested with our paradigm. On the right we can see the control group that use the old learning methods. In the naturalist intelligence (Fig. 11) we can see that the samples on the left side are more stable than the in the right one. This means that the acquired learning are much more stable using our method than the tradicional.

Test Group										Control Group									
@Samples	idGT01@%	idGT02@%	idGT03@%	idGT04@%	idGT05@%	idGT06@%	idGT07@%	idGT08@%	n/a	@Samples	idGC01@%	idGC02@%	idGC03@%	idGC04@%	idGC05@%	idGC06@%	idGC07@%	idGC08@%	idGC09@%
1	95.0	83.3	95.0	83.3	78.3	78.3	88.3	79.9		1	76.6	78.0	95.0	83.3	100.0	88.3	86.6	71.6	66.6
2	73.3	75.0	73.3	66.6	64.9	51.5	64.9	56.5		2	86.0	48.0	50.5	34.0	34.0	52.0	43.0	10.0	10.0
3	68.2	66.5	85.0	53.2	64.9	51.5	53.2	58.2		3	86.6	69.0	86.4	72.3	68.1	86.4	83.0	81.6	29.6
4	71.8	71.0	90.2	80.5	74.4	92.6	92.6	79.4		4	72.8	79.0	79.1	83.0	81.8	54.8	89.8	54.8	26.4
5	90.0	86.0	89.8	67.6	92.5	95.0	86.0	97.5		5	74.2	79.0	89.8	71.2	72.2	86.7	89.8	54.8	26.4
6	90.1	68.1	89.8	67.6	85.8	95.0	64.7	84.2		6	82	83	88.8	82.2	72.2	86.7	72.5	62	28.8
7	92.5	86	89.8	76.6	85.8	95	65.2	81.6		7	76.6	80	90.1	74.2	69.6	69.3	72.1	69.8	26.5
8	87.5	86.5	93.2	77.1	87.7	95	86	84.6		8	79	78	90.2	75	71	88.6	84.1	65.8	31.4
9	97.5	90.5	95	81.6	91.1	96.5	90	79.4		9	86	71	92	76	72.2	69.3	70.2	68.2	31.4
10	92	93	96	79.6	91.1	96	86.5	90.3		10	88	86	91	79	75.8	80.2	93.1	67.1	29.6

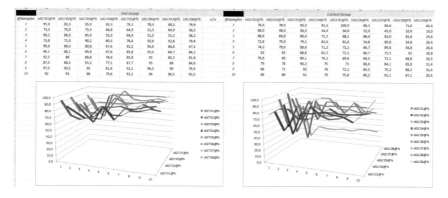

Fig. 11 Data@Animals

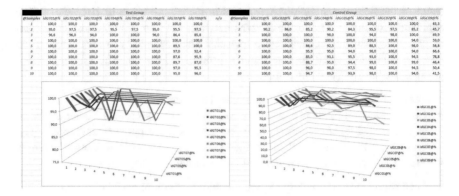

Fig. 12 Data@Geometric figures

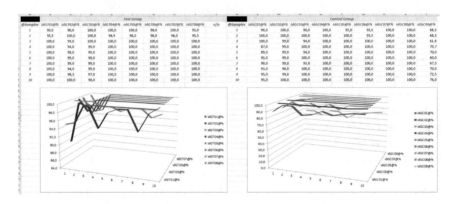

Fig. 13 Data@Paterns

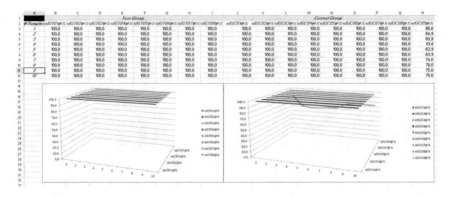

Fig. 14 Data@Words finished in ão

Regarding to the logical/mathematical intelligence we can observe that, again, the test group values are more stable than the control group.

In the visual-spatial intelligence data, both groups present similar data. We can conclude that, according to the cognitive theories, we must improve the way we transmit the visual-spatial knowledge through the KB structure.

In all the graphics (Figs. 11, 12, 13 and 14) we can saw that in the right side there are one subject—that we denominate as subject 9—that presents even worst results than the average values of the group. This children have profound educational learning gaps.

6.3 Phase III

Since the beginning of this investigation we collect several data from several fields —Education, Pedagogy, Psychology, and Human Relations, Motivation and others, to understand the complexity of the teaching/learning/acquiring knowledge process. The advantages of the proposed system are several, namely (summarizing):

- Accessible and immediate
- Interactive
- Resource reusing
- Dynamic and Cognition Oriented
- Context-awareness

With the application of this paradigm we can observe the following positive changes in the school context (summarizing):

- Teacher motivation in applying in a dynamic way the knowledge in the classroom
- Teacher motivation with the possibility of reuse the learning objects, and the facility of adapting old resources to the new model
- Understand some patterns in the learning process to better understand the tendencies of each student
- Allow the participation of parents in the educative process

The last phase of this research regarding to apply our method in special educational context was to prove that the method was dynamic enough to support his application into different knowledge fields. The results presented next, reports to the (Subject 9) data values.

In previous Fig. 15 we can observe that the Subject 9 has a serious educational issue. In eight matrix observations ($10 \times 10 \times 8$) he has a 10.39 (51.94%) average value, which is very low if compared with the average of the control group—18.09 (90.45%). This means a difference of 38.51% what represents a serious indicator of severe cognitive deficit. If we compare the Subject 9 to the test

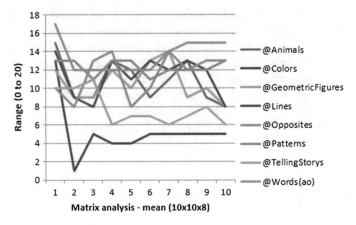

Fig. 15 "Subject 9" profile layout

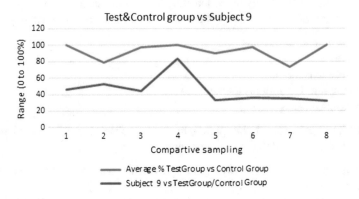

Fig. 16 Test and control groups versus "Subject 9"

group the difference raises to 42.06%—almost 4% difference more if compared with the average of the control group.

If compared, the means between the Test and Control groups versus Subject 9 (Fig. 16) it is visible the cognitive gap between them.

In Fig. 17 we can observe the future tendencies of both samplings. It is a linear prediction that show that the Subject 9 (red dotted line) tends to lower grades while the children's in the Test and Control groups (blue dotted line) tend to stabilize between 90 and 100%. This values represents the last proof that we need to validate our theory in the field of child's in special educational programs.

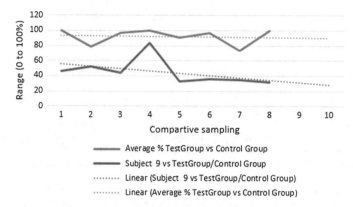

Fig. 17 Test and Control groups versus "Subject 9" (linear prediction)

7 Conclusion

This paper is an extended and improved version of the published in the Proceedings of the CISTI 2016 [23]. With this research we believe that a new educational paradigm was created. In this knowledge field a lot more it's possible to say, but we have always the time and space against us. Along the last 9 years the authors developed the educational paradigm and have apply it to normal educational curriculum and to special education programs. Using this methodology and the ubiquitous philosophy we can create the cognitive profile of people with disabilities, e.g. lack of vision, hearing.

The future work that authors intend to do is: adapt the system to other platforms and devices e.g. Android and IOS, improve the (AI) structure and extend the system to others disabilities, e.g. blind and deaf people. Because knowledge is property of everyone we think that even the so called "handicapped children" must have the best education. In this way our work has a meaning and is useful.

References

1. Salvador, P., Rocha, Á.: An assessment of content quality in websites of basic and secondary Portuguese schools. In: Rocha et al. (2014), New Perspectives in Information Systems and Technologies, vol. 1, pp. 71–82. Springer International Publishing (2014)
2. Abreu, A., Rocha, Á., Cota, M.P.: Perceptions of teachers and guardians on the electronic record in the school-family communication. In: Conference on e-Business, e-Services and e-Society, pp. 48–62. Springer International Publishing (2015, October)
3. Buzzi, M.C. et al.: Designing e-learning collaborative tools for blind people, Chapter 7. InTechOpen E-Learning—Long-Distance and Lifelong Perspectives (2012). ISBN: 978-953-51-0250-2, Published: 14 March 2012

4. Lino, A., Rocha, Á., Sizo, A.: Virtual teaching and learning environments: Automatic evaluation with symbolic regression. J. Intell. Fuzzy Syst. **31**(4), 2061–2072 (2016). doi:10. 3233/JIFS-169045
5. Luyi, L., Yanlin, Z., Fanglin, Z.: Design of a computer-supported ubiquitous learning system. School of Education Science, School of Media Science, Northeast Normal University, ChangChun, China
6. Hwang, G.-J.: Criteria and strategies of ubiquitous learning. Department of Information and Learning Technology, National University of Tainan, Taiwan
7. Kerkiri, T.A., et al.: A Learning Style—Driven Architecture Build on Open Source LMS's Infrastructure for Creation of Psycho-Pedagogically—'Savvy' Personalized Learning Paths, E-Learning Experiences and Future (2010). ISBN 978-953-307-092-6
8. Gardner, H.: Multiple Intelligences: The Theory in Practice. Basic Books. New York, USA (1993)
9. Feuerstein, R.: The theory of structural modifiability. In: Presseisen, B. (ed.), Learning and Thinking Styles: Classroom Interaction. National Education Associations. Washington DC, USA (1990)
10. Pires, J.M., Cota, M.P.: A new learning cognitive architecture using a statistical function and genetic algorithms—an intelligent new e-learning model. In: International Conference on e-Learning 2015 (eCONF '15), Manama, Bahrain (2015)
11. Grider, C.: Foudations of Cognitive Theory: A Concise Review. U.S: Department of Education, Office of Educational Research and Improvement, Educational Resources Information Center, Washington, USA (1993)
12. Sincero, S.M.: Cognitive Learning Theory. Retrieved from https://explorable.com/cognitive-learning-theory, 21 Sept. 2016
13. Madhour, H., Forte, M.W.: Personalized learning path delivery. In: Advances in Learning Processes (2010). ISBN 978-953-7619-56-5
14. Brusilovsky, P.: Adaptive and intelligent technologies for web-based education. Künstliche Intell. Spec. Issue Intell. Syst. Teleteach. **4**, 19–25 (1999)
15. Kazanidis, I., Satratzemi, M.: Towards the integration of adaptive educational systems with SCORM standard and authoring toolkits. In: Advanced Learning (2009). ISBN 978-953-307-010-0
16. Liu, F.-J., Shih, B.-J.: Application of data-mining technology on e-learning material recomendation. In: E-learning Experiences and Future (April 1, 2010). InTech, ISBN 978-953-307-092-6, Vukovar, Croatia
17. Groppo, M.A.: Doctoral Thesis—"Métodos de Evaluación por Computadoras para Reforzar la Interacción Docente-Alumno". University of Vigo, Spain (2010)
18. Pires, J.M.: Doctoral Thesis—"Evolutionary Intelligent E-Learning Applications Systems based on Genetic Algorithms", University of Vigo, Spain (January 18, 2016)
19. Skinner, B.F.: The Technology of Teaching. Appleton-Century-Crofts Publishers Inc, New York, USA (1968)
20. Piaget, J., Inhelder, B.: The of the Child, (1969). Basic Books Publishers Inc. ISBN 0-465-09500-3, New York, USA
21. Gagné, R.M.: Las condiciones del aprendizaje. Ediciones Aguilar, Madrid, Spain (1971)
22. Spiegel, M.R.: Estatística, MAKRON Books do Brasil Editora Lda, 3ª Edição, São Paulo, Brasil (1994)
23. Pires, J.M., Cota, M.P.: "Inteligent" adaptive learning objects applied to special education needs. Extending the eLearning Paradigm to the uLearning environment. In: 2016 11th Iberian Conference on Information Systems and Technologies (CISTI), pp. 1–6. IEEE (2016, June)

On Feature Weighting and Selection for Medical Document Classification

Bekir Parlak and Alper Kursat Uysal

Abstract Medical document classification is still one of the popular research problems inside text classification domain. In this study, the impact of feature selection and feature weighting on medical document classification is analyzed using two datasets containing MEDLINE documents. The performances of two different feature selection methods namely Gini index and distinguishing feature selector and two different term weighting methods namely term frequency (TF) and term frequency-inverse document frequency (TF-IDF) are analyzed using two pattern classifiers. These pattern classifiers are Bayesian network and C4.5 decision tree. As this study deals with single-label classification, a subset of documents inside OHSUMED and a self-constructed dataset is used for assessment of these methods. Due to having low amount of documents for some categories in self-compiled dataset, only documents belonging to 10 different disease categories are used in the experiments for both datasets. Experimental results show that the better result is obtained with combination of distinguishing feature selector, TF feature weighting, and Bayesian network classifier.

Keywords Text classification · Medical documents · Disease classification · MeSH

1 Introduction

Highly increase in the usage of Internet technology caused a significant growth in the number of electronic documents worldwide. This increase make automatic text classification approaches quite important. The main task of automatic text

B. Parlak (✉) · A.K. Uysal
Department of Computer Engineering, Anadolu University, Eskisehir, Turkey
e-mail: bekirparlak@anadolu.edu.tr

A.K. Uysal
e-mail: akuysal@anadolu.edu.tr

© Springer International Publishing AG 2018   269
Á. Rocha and L.P. Reis (eds.), *Developments and Advances in Intelligent Systems and Applications*, Studies in Computational Intelligence 718,
DOI 10.1007/978-3-319-58965-7_19

classification approach is to assign the electronic documents to the appropriate classes according to their content [1]. These documents can be retrieved from many different domains. It should be noted that every domain may have slightly different problems and solutions due to its nature. Text classification can be used to solve a variety of problems such as the filtering of spam e-mails [2], author identification [3], classification of web pages [4], sentiment classification [5, 6] and classification of medical text documents [7, 8–9].

Classification of medical abstracts is one of the main concerns inside medical text classification research field. Researches related to medical abstracts are generally carried out on MEDLINE database [10]. MEDLINE is a bibliographic database containing over 21 million documents, about 5600 medical journals. This database consists of medical abstracts in English which are assigned to some categories namely medical subject headings (MeSH). This database can be queried on internet through a search platform called PubMed [11]. Documents in MEDLINE database is indexed with corresponding relevant categories of MeSH terms by experts manually. In the literature, there exist some studies conducted on automatic classification of MEDLINE documents [8, 9, 12–21]. In these studies, datasets containing a certain amount of MEDLINE documents are used. The most used dataset for automatic classification of MEDLINE documents is called Ohsumed dataset. It contains medical abstracts in English for 23 types of diseases. Ohsumed, due to the structure of the MEDLINE database, is multi-label. So, it is necessary to apply multi-label classification approaches whenever a study on this dataset is performed using all documents.

In a previous study, the usage of words, medical phrases, and their combinations as features is investigated [8] for medical document classification. The results show that using combination of words and phrases as features gives slightly better classification performances than the others. In another study, multi-label classification performance based on associative classifier is examined on medical articles [12]. In another study, hidden Markov models are used for classification [16]. Besides, there exist a number of studies in the literature that ontology-based classification approaches are applied [14, 18]. In a recent study, an approach using support vector machines and latent semantic indexing is applied to some datasets including the ones consisting of medical abstracts [20]. Moreover, the performances of classifiers on medical document classification is analyzed for two cases where stemming is applied and not applied [21]. Also, the impact of different text representations of biomedical texts on the performance of classification are analyzed [9]. In a recent study [22], several experiments have been conducted using OHSUMED corpus. They obtained results using biomedical text categorization system based on three machine learning models. These models are support vector machine (SVM), naïve Bayes (NB) and maximum entropy (ME). The results show that the context-based methods (SenseRelate and NoDistanceSenseRelate) outperform the others. As a part of another study [23], a collection consisting of 1499

PubMed abstracts annotated according to the scientific evidence are used. They provide for the 10 currently known hallmarks of cancer to train a system that classifies PubMed literature according to the hallmarks. The system uses supervised machine learning and rich features largely based on biomedical text mining. In another study [24], the authors designed and assessed a method for extracting clinically useful sentences from synthesized online clinical resources that represent the most clinically useful information for directly answering clinicians' information needs. The feature-rich approach significantly outperformed general baseline methods. This approach significantly outperformed classifiers based on a single type of feature. Within the scope of one of the recent studies [25], the impact of feature selection on medical document classification is analyzed using two datasets containing MEDLINE documents. Gini index and distinguishing feature selector are used as two different feature selection methods. Two different pattern classifiers namely Bayesian network and C4.5 decision tree are utilized. As this study deals with single-label classification, a subset of documents inside OHSUMED and a self-constructed dataset is used for assessment of feature selection methods. According to experimental results, the combination of distinguishing feature selector and Bayesian network classifier gives more successful results in most cases than the others.

Apart from studies that uses MEDLINE documents, there exist some medical text classification studies using data obtained from various clinics data [13, 26–31]. Some of these studies concerns with medical text documents in different languages such as German [13].

In this study, the performances of two widely-known classifiers namely Bayesian networks and C4.5 decision trees are extensively analyzed using two feature selection methods on two different datasets consisting of MEDLINE documents. Also, a comparison on two different widely-known feature weighting methods is carried out in order to obtain the best combination of various parameters such as feature selection methods, feature weighting algorithms, and classifiers for medical document classification. In order to make a generalization from the results, two datasets having different characteristics are used in the experiments. The first dataset is a subset of well-known OHSUMED dataset. The second one is a self-constructed dataset whose data is retrieved programmatically with querying Pubmed search platform. This dataset differs from the first one. It consists of MEDLINE documents originated from medical journals in Turkey. However, it has smaller amount of data than the first dataset.

Rest of the paper is organized as follows: feature extraction and selection approaches used in the study are briefly described in Sect. 2. Section 3 explains pattern classifiers used in this study. Section 4 presents the experimental study and results. Finally, some concluding remarks are given in Sect. 5.

2 Feature Extraction and Selection

2.1 Feature Extraction

As in most of the text classification studies, bag of words approach [1, 21] can be used for feature extraction process. In this approach, the order of terms within documents is ignored and their occurrence frequencies are used [32, 39]. Therefore, each of the unique words in a text collection is considered as a different feature. Consequently, a document is represented by a multi-dimensional feature vector [1]. In a feature vector, each dimension corresponds to a value which is weighted by term frequency (TF), term frequency-inverse document frequency (TF-IDF), and etc. [33].

It should also be noted that it is necessary to apply some preprocessing steps during feature extraction from text documents. Widely used preprocessing steps are "stopword removal" and "stemming". In this study, both of these two steps were applied. Porter stemming algorithm [34] was used for stemming and two different term weighting approaches are applied. These two weighting approaches are TF and TF-IDF, respectively.

2.2 Feature Selection

Feature selection techniques generally fall into three categories: filters, wrappers, and embedded methods. Filter techniques are computationally fast; however, they usually do not take feature dependencies into consideration [1]. Filter-based methods are widely preferred especially for text classification domain. There is a mass amount of filter-based techniques for the selection of distinctive features in text classification. In this study, two different filter-based feature selection methods namely Gini index (GI) and distinguishing feature selector (DFS) were used. These methods are explained below in details.

2.2.1 Gini Index (GI)

GI is an improved version of the method originally used to find the best split of features in decision trees [35]. It is an accurate and fast method. Its formula is as below:

$$GI(t) = \sum_{i=1}^{M} P(t|C_i)^2 \cdot P(C_i|t)^2 \tag{1}$$

where $P(t|C_i)$ is the probability of term t given presence of class C_i, $P(C_i|t)$ is the probability of class C_i given presence of term t, respectively.

2.2.2 Distinguishing Feature Selector (DFS)

DFS is one of the recent successful feature selection methods for text classification [1] whose aim is to select distinctive features while eliminating uninformative ones considering some pre-determined criteria. DFS can be expressed with the following formula:

$$DFS(t) = \sum_{i=1}^{M} \frac{P(C_i|t)}{P(\bar{t}|C_i) + P(t|\bar{C_i}) + 1} \tag{2}$$

where M is the total number of classes, $P(C_i|t)$ is the conditional probability of class C_i given presence of term t, $P(\bar{t}|C_i)$ is the conditional probability of absence of term t given class C_i, and $P(t|\bar{C_i})$ is the conditional probability of term t given all the classes except C_i.

3 Pattern Classifiers

In this study, two classifiers in Weka [36] package were used programmatically. These are Bayesian Networks and C4.5 decision tree classifiers. These algorithms are explained in details below.

3.1 Bayesian Networks (BN)

BN is one of the methods which are used to denote modeling and state transitions [37]. BN is often used for modeling discrete and continuous variables of multi-nomial data. These networks encrypt the relationships between variables in the modeled data. In BN, the nodes are interconnected by arrows to indicate the direction of engagement with each other.

3.2 C4.5 Decision Tree (DT)

The main purpose of the decision tree algorithms is to split the feature space into unique regions corresponding to the classes [1]. An unknown feature vector is assigned to a class via a sequence of Yes/No decisions along a path of nodes of a decision tree. C4.5 is an algorithm used to generate a decision tree and it is known as one of the successful decision tree classification algorithms.

4   Experimental Work

In this section, an in-depth investigation was carried out to measure the performance of feature selection methods, term weighting methods and classifiers. For this purpose, combinations of feature selection methods with BN and DT classifiers were analyzed in order to determine the best combination for both of the datasets. At the same time, two different term weighting methods which are TF and TF-IDF are used. Also, the effect of dimension reduction can be inferred according to the experimental results. In the following subsections, the utilized datasets and success measures are briefly described. Then, the experimental results are presented.

4.1   Datasets

In this study, two different datasets containing MEDLINE documents were used. The first one is a subset of well-known Ohsumed dataset. It consists of medical abstracts collected in 1991 related to 23 cardiovascular disease categories. As this study deals with single-label text classification, the documents belonging to multiple categories are eliminated. Also, only 10 classes are used for classification in order to make the class distribution same with the second dataset. The second dataset is a self-constructed dataset whose data is retrieved programmatically with querying Pubmed search platform. This dataset is constructed via retrieving XML results containing medical abstracts and parsing it appropriately. The documents having multiple categories are removed from this dataset because of concerning single-label classification of medical documents. This dataset differs from the first one depending on its origins. It consists of MEDLINE documents only originated from medical journals in Turkey rather than originating from different locations. However, it has same categories with smaller amount of data than the first one. In this dataset, 10 categories having enough number of documents were used for the

Table 1 Ohsumed dataset

Class number	Disesase category	Number of documents
1	Bacterial infections and mycoses	631
2	Virus diseases	249
3	Parasitic diseases	183
4	Neoplasms	2513
5	Musculoskeletal diseases	505
7	Stomatognathic diseases	132
8	Respiratory tract diseases	634
10	Nervous system diseases	1328
14	Cardiovascular diseases	2876
23	Pathological conditions, signs and symptoms	1924

Table 2 Self-constructed dataset

Class number	Disesase category	Number of documents
1	Bacterial infections and mycoses	284
2	Virus diseases	44
3	Parasitic diseases	116
4	Neoplasms	32
5	Musculoskeletal diseases	140
7	Stomatognathic diseases	39
8	Respiratory tract diseases	90
10	Nervous system diseases	83
14	Cardiovascular diseases	231
23	Pathological conditions, signs and symptoms	73

evaluation. The detailed information regarding those datasets is provided in Tables 1 and 2. In the experiments, 70% of documents in each class was used for training. The rest was also used for testing.

4.2 Accuracy Analysis

Varying numbers of the features, which are selected by each selection method, were fed into DT and BN classifiers. In the experiments, stopword removal and stemming were applied. Widely-known Porter stemmer was carried out as stemming algorithm. In this study, GI and DFS are used as feature selection methods. Dimension reduction was carried out by constructing feature sets consisting of 300, 500, 1000, and 2000 features. Also, F-score [38] was used as success measure. This score is presented as both class specific and weighted averaged. Resulting F-Scores obtained on two datasets using TF and TF-IDF weighting approaches are listed in Tables 3, 4 and Tables 5, 6, respectively. The best ones in the results are shown as bolded.

Considering the highest weighted averaged F-scores, in most cases, DFS is superior to GI. In a small part of experiments, DFS and GI give similar results on both of the two datasets. It should be noted that DFS seems more successful when the feature size is low. Also, the scores obtained with TF weighting is generally more successful than the ones obtained with TF-IDF term weighting. In a small part of experiments, TF-IDF weighting is superior to TF weighting. It is common that TF and TF-IDF term weighting methods are both successful when the feature size is high. Besides, in spite of originated from different sources and having different class-based distributions, the maximum classification performances obtained on these two datasets are similar. BN classifier is more successful than DT classifier in most of the cases.

Table 3 Results on Ohsumed dataset (tf-weighted)

Number of features	Options				
	DFS + DT	DFS + BN	GI + BN	GI + DT	Classes
300	0.57	**0.65**	0.63	0.46	C1
	0.62	0.56	0.50	0.55	C2
	0.69	**0.77**	0.76	0.62	C3
	0.83	**0.85**	0.83	0.81	C4
	0.50	**0.58**	0.50	0.42	C5
	0.35	**0.59**	0.58	0.17	C7
	0.59	**0.62**	0.61	0.52	C8
	0.65	**0.67**	0.65	0.57	C10
	0.86	**0.86**	0.84	0.84	C14
	0.45	**0.47**	0.44	0.38	C23
Weighted average	0.69	**0.71**	0.68	0.64	
500	0.55	**0.67**	0.66	0.51	C1
	0.58	0.52	0.53	0.50	C2
	0.69	0.74	**0.78**	0.70	C3
	0.84	**0.84**	0.82	0.80	C4
	0.46	**0.57**	**0.57**	0.44	C5
	0.24	0.56	**0.57**	0.32	C7
	0.62	**0.62**	0.60	0.48	C8
	0.66	**0.66**	0.65	0.58	C10
	0.85	**0.86**	0.84	0.82	C14
	0.44	**0.45**	**0.45**	0.41	C23
Weighted average	0.69	**0.70**	0.69	0.64	
1000	0.55	**0.72**	0.68	0.50	C1
	0.58	0.52	0.51	0.50	C2
	0.71	**0.73**	0.70	0.68	C3
	0.83	**0.83**	0.82	0.82	C4
	0.47	**0.58**	**0.58**	0.46	C5
	0.27	**0.55**	0.51	0.24	C7
	0.61	**0.63**	0.62	0.54	C8
	0.63	**0.7**	0.68	0.58	C10
	0.84	**0.86**	0.85	0.81	C14
	0.43	**0.47**	0.46	0.41	C23
Weighted average	0.68	**0.71**	0.70	0.64	
2000	0.51	**0.72**	**0.72**	0.50	C1
	0.61	0.5	0.5	0.56	C2
	0.67	**0.74**	0.73	0.65	C3
	0.82	**0.84**	0.83	0.81	C4
	0.46	0.57	**0.58**	0.46	C5
	0.14	0.51	**0.52**	0.24	C7
	0.61	0.62	**0.63**	0.53	C8
	0.63	**0.71**	0.7	0.64	C10
	0.84	**0.86**	0.85	0.83	C14
	0.42	**0.47**	**0.47**	0.40	C23
Weighted average	0.67	**0.72**	0.71	0.65	

Table 4 Results on self-constructed dataset (tf-weighted)

Number of features	Options				
	DFS + DT	DFS + BN	GI + BN	GI + DT	Classes
300	**0.81**	**0.81**	0.79	0.8	C1
	0.67	0.42	0.44	0.57	C2
	0.86	0.72	0.72	0.84	C3
	0.63	0.31	0.31	0.57	C4
	0.62	0.76	**0.77**	0.68	C5
	0.67	**0.7**	**0.7**	0.67	C7
	0.74	0.55	0.59	0.6	C8
	0.27	0.43	**0.39**	0.3	C10
	0.72	**0.88**	0.87	0.69	C14
	0.53	0.49	0.52	**0.56**	C23
Weighted average	**0.70**	**0.70**	0.69	0.68	
500	0.8	**0.81**	**0.81**	**0.81**	C1
	0.59	0.31	0.31	**0.62**	C2
	0.84	0.77	0.77	**0.88**	C3
	0.63	0.31	0.31	**0.63**	C4
	0.67	**0.77**	**0.77**	0.71	C5
	0.67	**0.75**	**0.75**	0.67	C7
	0.63	0.58	0.57	0.56	C8
	0.37	0.45	**0.46**	0.36	C10
	0.67	**0.89**	0.88	0.7	C14
	0.57	0.53	0.53	**0.59**	C23
Weighted average	0.68	**0.71**	**0.71**	0.70	
1000	**0.82**	0.81	0.81	0.8	C1
	0.58	0.31	0.31	0.54	C2
	0.83	0.77	0.77	**0.85**	C3
	0.5	0.31	0.31	**0.57**	C4
	0.73	0.77	**0.77**	0.71	C5
	0.67	**0.75**	**0.75**	0.6	C7
	0.67	0.58	0.58	0.65	C8
	0.51	0.45	0.45	**0.51**	C10
	0.7	**0.89**	**0.89**	0.73	C14
	0.59	0.53	0.53	0.56	C23
Weighted average	**0.71**	**0.71**	**0.71**	**0.71**	
2000	**0.82**	0.81	0.81	0.8	C1
	0.54	0.31	0.31	**0.56**	C2
	0.87	0.77	0.77	**0.93**	C3
	0.67	0.43	0.31	0.63	C4
	0.69	**0.78**	0.77	0.74	C5
	0.63	**0.78**	0.75	0.67	C7
	0.58	0.57	0.58	**0.59**	C8
	0.58	0.46	0.45	0.46	C10
	0.73	**0.89**	**0.89**	0.68	C14
	0.39	**0.53**	**0.53**	0.5	C23
Weighted average	0.70	**0.71**	**0.71**	0.70	

Table 5 Results on Ohsumed dataset (tf-idf weighted)

Number of features	Options				
	DFS + DT	DFS + BN	GI + BN	GI + DT	Classes
300	0.57	**0.65**	0.63	0.43	C1
	0.61	**0.56**	0.50	0.52	C2
	0.69	**0.77**	0.76	0.62	C3
	0.83	**0.85**	0.83	0.80	C4
	0.49	**0.58**	0.50	0.41	C5
	0.35	**0.59**	0.58	0.19	C7
	0.59	**0.62**	0.61	0.52	C8
	0.64	**0.67**	0.65	0.56	C10
	0.85	**0.86**	0.84	0.84	C14
	0.45	**0.47**	0.44	0.37	C23
Weighted average	0.68	**0.71**	0.68	0.63	
500	0.55	**0.67**	0.66	0.48	C1
	0.58	0.52	**0.53**	0.45	C2
	0.67	0.74	**0.78**	0.69	C3
	0.83	**0.84**	0.82	0.79	C4
	0.45	**0.57**	**0.57**	0.41	C5
	0.24	0.56	**0.57**	0.30	C7
	0.63	**0.62**	0.60	0.47	C8
	0.66	**0.66**	0.65	0.55	C10
	0.84	**0.86**	0.84	0.81	C14
	0.44	**0.45**	**0.45**	0.39	C23
Weighted average	0.68	**0.70**	0.69	0.63	
1000	0.54	**0.72**	0.68	0.50	C1
	0.58	**0.52**	0.51	0.51	C2
	0.69	**0.73**	0.70	0.68	C3
	0.83	**0.83**	0.82	0.81	C4
	0.48	**0.58**	0.58	0.44	C5
	0.30	**0.55**	0.51	0.25	C7
	0.61	**0.63**	0.62	0.53	C8
	0.62	**0.70**	0.68	0.56	C10
	0.84	**0.86**	0.85	0.80	C14
	0.42	**0.47**	0.46	0.39	C23
Weighted average	0.67	**0.71**	0.70	0.63	
2000	0.51	0.72	**0.81**	0.80	C1
	0.59	0.50	**0.31**	0.61	C2
	0.67	0.74	**0.77**	0.90	C3
	0.82	0.84	**0.31**	0.63	C4
	0.46	0.57	**0.77**	0.74	C5
	0.14	0.51	**0.75**	0.67	C7
	0.60	0.62	**0.58**	0.64	C8
	0.63	0.71	**0.45**	0.50	C10
	0.84	0.86	**0.89**	0.69	C14
	0.42	0.47	**0.53**	0.47	C23
Weighted average	0.67	**0.71**	0.71	0.71	

Table 6 Results on self-constructed dataset (tf-idf weighted)

Number of features	Options				
	DFS + DT	DFS + BN	GI + BN	GI + DT	Classes
300	0.79	**0.78**	0.78	0.77	C1
	0.64	**0.46**	0.48	0.56	C2
	0.84	**0.72**	0.70	0.86	C3
	0.60	**0.47**	0.53	0.46	C4
	0.69	**0.79**	0.78	0.68	C5
	0.67	**0.76**	0.67	0.67	C7
	0.55	**0.57**	0.46	0.52	C8
	0.27	**0.39**	0.24	0.16	C10
	0.69	**0.89**	0.85	0.64	C14
	0.45	**0.49**	0.43	0.44	C23
Weighted average	**0.67**	**0.69**	0.66	0.64	
500	0.80	0.78	0.78	**0.79**	C1
	0.59	0.25	0.37	**0.54**	C2
	0.84	0.70	0.68	**0.88**	C3
	0.44	0.37	0.50	**0.46**	C4
	0.72	0.78	0.79	**0.70**	C5
	0.67	0.76	0.76	**0.63**	C7
	0.58	0.57	0.54	**0.63**	C8
	0.37	0.39	0.38	**0.26**	C10
	0.69	0.89	0.86	**0.66**	C14
	0.54	0.52	0.46	**0.40**	C23
Weighted average	0.68	**0.68**	**0.68**	0.66	
1000	**0.82**	0.78	0.78	0.80	C1
	0.64	0.25	0.32	0.50	C2
	0.89	0.70	0.70	0.86	C3
	0.50	0.40	0.43	0.44	C4
	0.62	0.79	0.81	0.71	C5
	0.67	0.76	0.76	0.67	C7
	0.67	0.56	0.57	0.70	C8
	0.47	0.41	0.39	0.52	C10
	0.68	0.90	0.88	0.73	C14
	0.53	0.53	0.50	0.55	C23
Weighted average	0.70	0.69	0.69	**0.71**	
2000	0.82	**0.78**	**0.78**	0.82	C1
	0.54	**0.25**	**0.32**	0.58	C2
	0.84	**0.70**	**0.70**	0.87	C3
	0.67	**0.40**	**0.43**	0.52	C4
	0.71	**0.79**	**0.81**	0.74	C5
	0.67	**0.76**	**0.76**	0.67	C7
	0.59	**0.56**	**0.57**	0.56	C8
	0.54	**0.41**	**0.39**	0.39	C10
	0.69	**0.90**	**0.88**	0.67	C14
	0.38	**0.53**	**0.50**	0.50	C23
Weighted average	0.69	**0.69**	**0.69**	0.69	

Considering class based F-scores, classification performances obtained on neoplasms (C4) and cardiovascular diseases (C14) categories are generally higher than the others for the first dataset. The results are unchanged when applying two different term weighting methods which are TF and TF-IDF methods in two datasets. This may be due to having high amount of training instances for these two categories. For self-constructed dataset, classification performances obtained on parasitic diseases (C3) and cardiovascular diseases (C14) categories are generally higher than the others. TF and TF-IDF term weighting methods did not change the results both Ohsumed and self-constructed dataset. In this case, these are not the classes with maximum number of documents. This situation may be caused by having small amount of data for most of the categories. Also, for most of the class-based F-scores, combination of DFS and BN seems better than the other ones.

5   Conclusions

In this study, the performances of two widely-known classifiers are extensively analyzed using two different feature selection methods. Two different term weighting methods are also used in the experiments. This analysis is realized on two different datasets consisting of MEDLINE documents. In the experiments, stopword removal and stemming as preprocessing steps are applied. Experimental results show that the most successful setting is the combination of Bayesian Network classifier, distinguishing feature selector, and TF term weighting method. As a future work, a new dataset containing Turkish versions of the documents in the self-constructed dataset may be compiled and classification performances of these two datasets having same documents in different languages can be extensively analyzed. In this paper, we have revised and extended the research results presented earlier in [25].

Acknowledgements This work was supported by Anadolu University, Fund of Scientific Research Projects under grant number 1503F136.

References

1. Uysal, A.K., Gunal, S.: A novel probabilistic feature selection method for text classification. Knowl.-Based Syst. **36**, 226–235 (2012)
2. Idris, I., Selamat, A., Nguyen, N.T., Omatu, S., Krejcar, O., Kuca, K., Penhaker, M.: A combined negative selection algorithm—particle swarm optimization for an email spam detection system. Eng. Appl. Artif. Intell. **39**, 33–44 (2015)
3. Zhang, C., Wu, X., Niu, Z., Ding, W.: Authorship identification from unstructured texts. Knowl.-Based Syst. **66**, 99–111 (2014)
4. Ozel, S.A.: A Web page classification system based on a genetic algorithm using tagged-terms as features. Expert Syst. Appl. **38**(4), 3407–3415 (2011)

5. Agarwal, B., Mittal, N.: Prominent Feature Extraction for Sentiment Analysis, pp. 21–45. Springer (2016)
6. Pak, M.Y., Gunal, S.: Sentiment classification based on domain prediction. Elektronika ir Elektrotechnika **22**(2), 96–99 (2016)
7. Garla, V., Taylor, C., Brandt, C.: Semi-supervised clinical text classification with Laplacian SVMs: an application to cancer case management. J. Biomed. Inform. **46**(5), 869–875 (2013)
8. Yetisgen-Yildiz, M., Pratt, W.: The effect of feature representation on MEDLINE document classification. In: AMIA Annual Symposium Proceedings, p. 849. American Medical Informatics Association (2005)
9. Yepes, A.J.J., Plaza, L., Carrillo-de-Albornoz, J., Mork, J.G., Aronson, A.R.: Feature engineering for MEDLINE citation categorization with MeSH. BMC Bioinform. **16**(1), 1 (2015)
10. MEDLINE. [http://www.nlm.nih.gov/databases/databases_medline.html]. Accessed 2015
11. Pubmed [http://www.ncbi.nlm.nih.gov/pubmed]. Accessed 2015
12. Rak, R., Kurgan, L.A., Reformat, M.: Multilabel associative classification categorization of MEDLINE articles into MeSH keywords. IEEE Eng. Med. Biol. Mag. **26**(2), 47 (2007)
13. Spat, S., Cadonna, B., Rakovac, I., Gutl, C., Leitner, H., Stark, G., Beck, P.: Multi-label text classification of German language medical documents. In: Proceedings of the 12th World Congress on Health (Medical) Informatics; Building Sustainable Health Systems, p. 2343 (2007)
14. Camous, F., Blott, S., Smeaton, A.F.: Ontology-based MEDLINE document classification. In: Bioinformatics Research and Development, pp. 439–452. Springer Berlin Heidelberg (2007)
15. Poulter, G.L., Rubin, D.L., Altman, R.B.: Seoighe, C.: MScanner: a classifier for retrieving medline citations. BMC Bioinform. **9**(1), 108 (2008)
16. Yi, K., Beheshti, J.: A hidden Markov model-based text classification of medical documents. J. Inf. Sci. (2008)
17. Frunza, O., Inkpen, D., Matwin, S., Klement, W., O'blenis, P.: Exploiting the systematic review protocol for classification of medical abstracts. Artif. Intell. Med. **51**(1), 17–25 (2011)
18. Dollah, R.B., Aono, M.: Ontology based approach for classifying biomedical text abstracts. Int. J. Data Engi. (IJDE), **2**(1), 1–15 (2011)
19. Albitar, S., Espinasse, B., Fournier, S.: Semantic enrichments in text supervised classification: application to medical domain. In: The Twenty-Seventh International Flairs Conference (2014)
20. Uysal, A.K., Gunal, S.: Text classification using genetic algorithm oriented latent semantic features. Expert Syst. Appl. **41**(13), 5938–5947 (2014)
21. Parlak, B., Uysal, A. K.: Classification of medical documents according to diseases. In: 23th IEEE Signal Processing and Communications Applications Conference (SIU), pp. 1635–1638 (2015)
22. Rais, M., Lachkar, A.: Evaluation of disambiguation strategies on biomedical text categorization. In: International Conference on Bioinformatics and Biomedical Engineering, pp. 790–801. Springer International Publishing (2016)
23. Baker, S., Silins, I., Guo, Y., Ali, I., Högberg, J., Stenius, U., Korhonen, A.: Automatic semantic classification of scientific literature according to the hallmarks of cancer. Bioinformatics **32**(3), 432–440 (2016)
24. Morid, M.A., Fiszman, M., Raja, K., Jonnalagadda, S.R., Del Fiol, G.: Classification of clinically useful sentences in clinical evidence resources. J. Biomed. Inform. **60**, 14–22 (2016)
25. Parlak, B., Uysal, A.K.: The impact of feature selection on medical document classification. In: 11th Iberian Conference on Information Systems and Technologies (CISTI), pp. 1–5 (2016)
26. Pakhomov, S.V., Buntrock, J.D., Chute, C.G.: Automating the assignment of diagnosis codes to patient encounters using example-based and machine learning techniques. J. Am. Med. Inform. Assoc. **13**(5), 516–525 (2006)

27. Van Der Zwaan, J., Sang, E.T.K., de Rijke, M.: An experiment in automatic classification of pathological reports. In: Artificial Intelligence in Medicine, pp. 207–216. Springer, Berlin Heidelberg (2007)
28. Waraporn, P., Meesad, P., Clayton, G.: Ontology-supported processing of clinical text using medical knowledge integration for multi-label classification of diagnosis coding (2010). arXiv:1004.1230
29. Boytcheva, S.: Automatic matching of ICD-10 codes to diagnoses in discharge letters. In: Proceedings of the Workshop on Biomedical Natural Language Processing, pp. 11–18. Hissar, Bulgaria (2011)
30. Ceylan, N.M., Alpkocak, A., Esatoglu, A.E.: Tıbbi Kayıtlara ICD-10 Hastalık Kodlarının Atanmasına Yardımcı Akıllı Bir Sistem (2012)
31. Arifoglu, D., Deniz, O., Alecakır, K., Yondem, M.: CodeMagic: semi-automatic assignment of ICD-10-AM codes to patient records. In: Information Sciences and Systems 2014, pp. 259–268. Springer International Publishing (2014)
32. Uysal, A.K., Gunal, S., Ergin, S., Gunal, E.S.: Detection of SMS spam messages on mobile phones. In: 20th IEEE Signal Processing and Communications Applications Conference (SIU), pp. 1–4 (2012)
33. Manning, C.D., Raghavan, P., Schutze, H.: Introduction to Information Retrieval Cambridge University Press, New York, USA (2008)
34. Porter, M.F.: An algorithm for suffix stripping. Program **14**, 130–137 (1980)
35. Shang, W., Huang, H., Zhu, H., Lin, Y., Qu, Y., Wang, Z.: A novel feature selection algorithm for text categorization. Expert Syst. Appl. **33**(1), 1–5 (2007)
36. Hall, M., Frank, E., Holmes, G., Pfahringer, B., Reutemann, P., Witten, I.H.: The WEKA Data Mining Software: An Update. SIGKDD Explor. **11**(1) (2009)
37. Witten, I.H., Frank, E.: Data Mining: Practical Machine Learning Tools and Techniques, Jim Gray (ed.). Morgan Kaufmann Publishers, San Fransisco (2005)
38. Goutte, C., Gaussier, E.: A probabilistic interpretation of precision, recall and F-score, with implication for evaluation. In: Proceedings of the Europe Conference Information Retrieval Research, pp. 345–359 (2005)
39. Rocha, A., Rocha, B.: Adopting nursing health record standards. Inform. Health Soc. Care **39**(1), 1–14 (2014)

Printed in the United States
By Bookmasters